T0202724

Undergraduate Texts in Mathematics

Undergraduate Texts in Mathematics

Undergraduate Texts in Mathematics are generally aimed at third- and fourth-year undergraduate mathematics students at North American universities. These texts strive to provide students and teachers with new perspectives and novel approaches. The books include motivation that guides the reader to an appreciation of interrelations among different aspects of the subject. They feature examples that illustrate key concepts as well as exercises that strengthen understanding.

For further volumes:
http://www.springer.com/series/666

Erhan Çınlar • Robert J. Vanderbei

Real and Convex Analysis

 Springer

Erhan Çınlar
Department of Operations Research
  and Financial Engineering
Princeton University
Princeton, New Jersey
USA

Robert J. Vanderbei
Department of Operations Research
  and Financial Engineering
Princeton University
Princeton, New Jersey
USA

ISSN 0172-6056
ISBN 978-1-4899-9859-0   ISBN 978-1-4614-5257-7 (eBook)
DOI 10.1007/978-1-4614-5257-7
Springer New York Heidelberg Dordrecht London

Mathematics Subject Classification: MSC 2010: 97I10, 26-01, 28-01, 34-01, 40-01, 49-01, 52-01
© Springer Science+Business Media New York 2013
Softcover re-print of the Hardcover 1st edition 2013

Printed on acid-free paper

Springer is part of Springer Science+Business Media (www.springer.com)

Preface

This book is intended to serve as a first course in analysis for scientists and engineers. It can be used either at the advanced undergraduate level or as part of the curriculum in a graduate program. We have taught from preliminary drafts of the book for several years.

The book is built around metric spaces. In the first three chapters, we lay the foundational material. We cover the all-important "four Cs": convergence, completeness, compactness, and continuity. We have organized the material to be as simple and as logical as possible.

In subsequent chapters, we use the basic tools of analysis to give a brief introduction to closely related topics such as differential and integral equations, convex analysis, and measure theory. The book is short and yet covers in some depth the most important subjects. We gave careful consideration to what to include and what to leave out. In all such considerations, we asked ourselves whether the material would be of direct and immediate use to scientists and engineers. Our philosophy is "if in doubt, do without."

What makes this book different? We pull together some of the foundational material one might find, for example, in the classic book by Rudin [Rud76] with material on convexity and optimization at a level commensurate, say, with the book by Borwein and Lewis [BL06] and with a completely modern treatment of the basics of measure theory. The importance of measure theory has increased over the years as stochastic modeling has become more central to all aspects of analysis. Similarly, optimization plays an ever increasing role as one tries to design and analyze the best possible "widget."

We hope that the reader will enjoy the book and learn some important mathematics.

We would like to thank the many students whom we have had the pleasure of teaching over the years. We give a special thanks to John D'Angelo; he carefully read a draft of the manuscript and made numerous helpful suggestions.

E. Çınlar and R.J. Vanderbei

Contents

Notation and Usage

We use the terms "positive" and "negative" in their wide sense: positive means ≥ 0, negative means ≤ 0. Similarly, "increasing" means $x \leq y$ implies $f(x) \leq f(y)$. If strict inequalities hold, we say "strictly positive," "strictly negative," "strictly increasing," etc. Here is a list of frequently used notations.

$\emptyset$: The *empty set*.

$\mathbb{N} = \{0, 1, 2, \dots\}$: The set of *natural numbers*.

$\mathbb{N}^* = \{1, 2, 3, \dots\}$: The set of *strictly positive integers*.

$\mathbb{Z} = \{0, 1, -1, 2, -2, \dots\}$: The set of *integers*.

$\mathbb{Q} = \{x : x = \frac{m}{n} \text{ for some } m \text{ in } \mathbb{Z} \text{ and some } n \text{ in } \mathbb{N}^*\}$: The set of *rationals*.

$\mathbb{R} = (-\infty, \infty) = \{x : -\infty < x < +\infty\}$: The set of *reals*.

$\mathbb{R}_+ = [0, \infty) = \{x \in \mathbb{R} : x \geq 0\}$: The set of *positive reals*.

$\mathbb{R}^* = (-\infty, \infty]$: The set of *reals and plus infinity*.

$\bar{\mathbb{R}} = [-\infty, \infty]$: The set of *extended reals*.

$[a, b] = \{x \in \mathbb{R} : a \leq x \leq b\}$: The *closed interval* with endpoints a and b.

$(a, b) = \{x \in \mathbb{R} : a < x < b\}$: The *open interval* with endpoints a and b.

$\log(x)$: The *natural logarithm* of x.

$x \cdot y = \sum_1^n x_i y_i$: The *inner product* of x and y.

$\|x\| = \sqrt{x \cdot x}$: The *Euclidean norm* of x.

$\mathcal{C}$: The set of *continuous functions*.

$d(x, y)$: The *distance* from x to y.

$B(x, r) = \{y : d(x, y) < r\}$: The *open ball* centered on x of radius r.

A°: The *interior* of the set A.

$\bar{A}$: The *closure* of the set A.

∂A: The *boundary* of the set A.

$\hat{f}$: The *Legendre transform* of the function f.

$f \star g$: The *infimal convolution* of the functions f and g.

$x \wedge y$: The *minimum* of the real numbers x and y.

$x \vee y$: The *maximum* of the real numbers x and y.

1_A: The *indicator function* of the set A.

$\mathcal{B}(E)$: The *Borel σ-algebra* on the metric space E.

CHAPTER 1

Sets and Functions

This introductory chapter is devoted to general notions regarding sets, functions, sequences, and series. We aim to introduce and review the basic notation, terminology, conventions, and elementary facts.

A. Sets

A *set* is a collection of some objects. Given a set, the objects that form it are called its *elements*. Given a set A, we write $x \in A$ to mean that x is an element of A. To say that $x \in A$, we also say x is in A, or x is a member of A, or x belongs to A, or A includes x.

One can specify a set by listing its elements inside curly braces, but doing so is not feasible in most cases. More often we specify a set by precisely describing its elements. For example, $A = \{a, b, c\}$ is the set whose elements are a, b, and c, and $B = \{x : x > 2.7\}$ is the set of all numbers exceeding 2.7. The following are some special sets:

$\emptyset$: The *empty set*. It has no elements.

$\mathbb{N} = \{0, 1, 2, \dots\}$: The set of *natural numbers*.

$\mathbb{N}^* = \{1, 2, 3, \dots\}$: The set of *strictly positive integers*.

$\mathbb{Z} = \{0, 1, -1, 2, -2, \dots\}$: The set of *integers*.

$\mathbb{Q} = \{x : x = \frac{m}{n} \text{ for some } m \text{ in } \mathbb{Z} \text{ and some } n \text{ in } \mathbb{N}^*\}$: The set of *rationals*.

$\mathbb{R} = (-\infty, \infty) = \{x : -\infty < x < +\infty\}$: The set of *reals*.

$\mathbb{R}_+ = [0, \infty) = \{x \in \mathbb{R} : x \geq 0\}$: The set of *positive reals*.

$[a, b] = \{x \in \mathbb{R} : a \leq x \leq b\}$: The *closed interval* with endpoints a and b.

$(a, b) = \{x \in \mathbb{R} : a < x < b\}$, defined for numbers a and b with $a < b$: The *open interval* with endpoints a and b.

We assume that the reader is familiar with these sets. For example, we take it for granted that real numbers are limits of rationals.

Subsets

A set A is said to be a *subset* of a set B if every element of A is an element of B. We write $A \subset B$ or $B \supset A$ to indicate this and say A is contained in B, or B contains A, to the same effect. The sets A and B are the same if and only if $A \subset B$ and $A \supset B$,

E. Çınlar and R.J. Vanderbei, *Real and Convex Analysis*, Undergraduate Texts in Mathematics, DOI 10.1007/978-1-4614-5257-7_1, © Springer Science+Business Media New York 2013

and then we write $A = B$. For the contrary situations, we write $A \neq B$ when A and B are not the same. The set A is called a *proper subset* of B if A is a subset of B, and A and B are not the same.

The empty set is a subset of every set. Let A be a set. The claim is that $\emptyset \subset A$, that is, that every element of $\emptyset$ is also an element of A, or equivalently, there is no element of $\emptyset$ that does not belong to A. But the last phrase is true simply because $\emptyset$ has no elements.

Set Operations

Let A and B be sets. Their *union*, denoted by $A \cup B$, is the set consisting of all elements that belong to either A or B (or both). Their *intersection*, denoted by $A \cap B$, is the set of all elements that belong to both A and B. The *complement* of A in B, denoted by $B \setminus A$, is the set of all elements of B that are not in A. Sometimes, when B is understood from context, $B \setminus A$ is also called the complement of A and is denoted by A^c. Regarding these operations, the following statements hold:

Commutative laws:

$$\begin{aligned} A \cup B &= B \cup A, \\ A \cap B &= B \cap A. \end{aligned}$$

Associative laws:

$$\begin{aligned} (A \cup B) \cup C &= A \cup (B \cup C), \\ (A \cap B) \cap C &= A \cap (B \cap C). \end{aligned}$$

Distributive laws:

$$\begin{aligned} A \cap (B \cup C) &= (A \cap B) \cup (A \cap C), \\ A \cup (B \cap C) &= (A \cup B) \cap (A \cup C). \end{aligned}$$

The associative laws show that $A \cup B \cup C$ and $A \cap B \cap C$ have unambiguous meanings.

Definitions of unions and intersections can be extended to arbitrary collections of sets. Let I be a set. For each i in I, let A_i be a set. The union of the sets A_i, $i \in I$, is the set A such that $x \in A$ if and only if $x \in A_i$ for some i in I. The intersection of them is the set A such that $x \in A$ if and only if $x \in A_i$ for every i in I. The following notations are used to denote the union and intersection respectively:

$$\bigcup_{i \in I} A_i, \qquad \bigcap_{i \in I} A_i.$$

When $I = \{1, 2, 3, \dots\}$, it is customary to write

$$\bigcup_{i=1}^{\infty} A_i, \qquad \bigcap_{i=1}^{\infty} A_i.$$

All of these notations follow the conventions for sums of numbers. For instance,

$$\bigcup_{i=1}^{n} A_i = A_1 \cup \cdots \cup A_n, \qquad \bigcap_{i=3}^{9} A_i = A_3 \cap A_4 \cap \cdots \cap A_9.$$

Disjoint Sets

Two sets are said to be *disjoint* if their intersection is empty; that is, if they have no elements in common. A collection $\{A_i : i \in I\}$ of sets is said to be *disjointed* if A_i and A_j are disjoint for every i and j in I with $i \neq j$.

Products of Sets

Let A and B be sets. Their *product*, denoted by $A \times B$, is the set of all pairs (x, y) with x in A and y in B. It is also called the *rectangle* with sides A and B.

If $A_1, \ldots, A_n$ are sets, then their product $A_1 \times \cdots \times A_n$ is the set of all n-tuples $(x_1, \ldots, x_n)$ where $x_1 \in A_1, \ldots, x_n \in A_n$. This product is called, variously, a rectangle, or a box, or an n-dimensional box. If $A_1 = \cdots = A_n = A$, then $A_1 \times \cdots \times A_n$ is denoted by A^n. Thus, $\mathbb{R}^2$ is the plane, $\mathbb{R}^3$ is the three-dimensional space, $\mathbb{R}_+^2$ is the positive quadrant of the plane, etc.

Exercises

1.1 Let E be a set. Show the following for subsets A, B, C, and A_i of E. Here, all complements are with respect to E; for instance, $A^c = E \setminus A$.

    (a) $(A^c)^c = A$
    (b) $B \setminus A = B \cap A^c$
    (c) $(B \setminus A) \cap C = (B \cap C) \setminus (A \cap C)$
    (d) $(A \cup B)^c = A^c \cap B^c$
    (e) $(A \cap B)^c = A^c \cup B^c$
    (f) $(\bigcup_{i \in I} A_i)^c = \bigcap_{i \in I} A_i^c$
    (g) $(\bigcap_{i \in I} A_i)^c = \bigcup_{i \in I} A_i^c$

1.2 Let a and b be real numbers with $a < b$. Find

$$\bigcup_{n=1}^{\infty} \left[a + \frac{1}{n}, b - \frac{1}{n}\right], \qquad \bigcap_{n=1}^{\infty} \left[a - \frac{1}{n}, b + \frac{1}{n}\right].$$

1.3 Describe the following sets in words and pictures:

    (a) $A = \{x \in \mathbb{R}^2 : x_1^2 + x_2^2 < 1\}$

   (b) $B = \{x \in \mathbb{R}^2 : x_1^2 + x_2^2 \leq 1\}$
   (c) $C = B \setminus A$
   (d) $D = C \times B$
   (e) $S = C \times C$

1.4   Let A_n be the set of points (x, y) in $\mathbb{R}^2$ lying on the curve $y = 1/x^n, 0 < x < \infty$. What is $\bigcap_{n \geq 1} A_n$?

B. Functions and Sequences

Let E and F be sets. With each element x of E, let there be associated a unique element $f(x)$ of F. Then f is called a *function* from E into F, and f is said to *map E into F*. We write $f \colon E \mapsto F$ to indicate this.

Let f be a function from E into F. For x in E, the point $f(x)$ in F is called the *image* of x or the value of f at x. Similarly, for $A \subset E$, the set

$$\{y \in F : y = f(x) \text{ for some } x \in A\}$$

is called the *image* of A. In particular, the image of E is called the *range* of f. Moving in the opposite direction, for $B \subset F$,

1.5 $$f^{-1}B = \{x \in E : f(x) \in B\}$$

is called the *inverse image* of B under f. Obviously, the inverse image of F is E.

Terms like mapping, operator, transformation are synonyms for the term "function" with varying shades of meaning depending on the context and on the sets E and F. We shall become familiar with them in time. Sometimes, we write $x \mapsto f(x)$ to indicate the mapping f; for instance, the mapping $x \mapsto x^3 + 5$ from $\mathbb{R}$ into $\mathbb{R}$ is the function $f \colon \mathbb{R} \mapsto \mathbb{R}$ defined by $f(x) = x^3 + 5$.

Injections, Surjections, Bijections

Let f be a function from E into F. It is called an *injection*, or is said to be *injective*, or is said to be *one-to-one*, if distinct points have distinct images, that is, if $x \neq y$ implies $f(x) \neq f(y)$. It is called a *surjection*, or is said to be *surjective*, if its range is F, in which case f is said to be from E *onto* F. It is called a *bijection*, or is said to be *bijective*, if it is both injective and surjective.

These terms are relative to E and F. For example, $x \mapsto e^x$ is an injection from $\mathbb{R}$ into $\mathbb{R}$, but is a bijection from $\mathbb{R}$ onto $(0, \infty)$. The function $x \mapsto \sin x$ from $\mathbb{R}$ into $\mathbb{R}$ is neither injective nor surjective, but it is a surjection from $\mathbb{R}$ onto $[-1, 1]$.

Sequences

A *sequence* is a function from $\mathbb{N}^*$ into some set. If f is a sequence, it is customary to denote $f(n)$ by something like x_n and write (x_n) or $(x_1, x_2, \dots)$ for the sequence (instead of f). Then, the x_n are called the *terms* of the sequence. For instance, $(1, 3, 4, 7, 11, \dots)$ is a sequence whose first, second, etc. terms are $x_1 = 1$, $x_2 = 3$, etc. Sometimes it is convenient to define a sequence over $\mathbb{N}$, and then write $(x_n)_{n \in \mathbb{N}}$ or $(x_0, x_1, \dots)$ for it.

If A is a set and every term of the sequence (x_n) belongs to A, then (x_n) is said to be a sequence in A or a sequence of elements of A, and we write $(x_n) \subset A$ to indicate this with a slight abuse of notation.

A sequence (x_n) is said to be a *subsequence* of (y_n) if there exist integers $1 \leq k_1 < k_2 < k_3 < \cdots$ such that

$$x_n = y_{k_n}$$

for each n. For instance, the sequence $(1, 1/2, 1/4, 1/8, \dots)$ is a subsequence of $(1, 1/2, 1/3, 1/4, 1/5, \dots)$.

Exercises

1.6 *Inverse images.* Let f be a mapping from E into F. Show that

    (a) $f^{-1}\emptyset = \emptyset$,
    (b) $f^{-1}F = E$,
    (c) $f^{-1}(B \setminus C) = (f^{-1}B) \setminus (f^{-1}C)$,
    (d) $f^{-1}(\bigcup_{i \in I} B_i) = \bigcup_{i \in I} f^{-1}B_i$,
    (e) $f^{-1}(\bigcap_{i \in I} B_i) = \bigcap_{i \in I} f^{-1}B_i$,

for all subsets B, C, B_i of F.

1.7 *Exponential and logarithm.* Show that $x \mapsto e^{-x}$ is a bijection from $\mathbb{R}_+$ onto $(0, 1]$. Show that $x \mapsto \log x$ is a bijection from $(0, \infty)$ onto $\mathbb{R}$. (Incidentally, $\log x$ is the logarithm of x to the base e, which is nowadays called the natural logarithm. We call it the logarithm. Let others call their logarithms "unnatural" and, while they are at it, they can also invent unnatural exponentials like $x \mapsto a^x$.)

1.8 *Bijections between* $\mathbb{N}^*$ *and* $\mathbb{Z}$. Let f be defined by the arrows below (for instance, $f(6) = -3$):

$$
\begin{array}{ccccccccc}
1 & 2 & 3 & 4 & 5 & 6 & 7 & \cdots \\
\downarrow & \downarrow & \downarrow & \downarrow & \downarrow & \downarrow & \downarrow & \\
0 & -1 & 1 & -2 & 2 & -3 & 3 & \cdots
\end{array}
$$

This defines a bijection from $\mathbb{N}^*$ onto $\mathbb{Z}$. Using this, construct a bijection from $\mathbb{Z}$ onto $\mathbb{N}^*$.

1.9 Bijection from $\mathbb{Z} \times \mathbb{Z}$ *onto* $\mathbb{N}^*$. Let $f \colon \mathbb{N}^* \times \mathbb{N}^* \mapsto \mathbb{N}^*$ be defined by the table below where $f(i,j)$ is the entry in the ith row and the jth column. Use this and the preceding exercise to construct a bijection from $\mathbb{Z} \times \mathbb{Z}$ onto $\mathbb{N}^*$.

$\diagdown$ $\quad j$ i $\quad \diagdown$	1	2	3	4	5	6	$\cdots$
1	1	3	6	10	15	21	
2	2	5	9	14	20		
3	4	8	13	19			
4	7	12	18				
5	11	17					
6	16						
$\vdots$							

1.10 Functional inverses. Let f be a bijection from E onto F. Then, for each y in F there is a unique x in E such that $f(x) = y$. In other words, in the notation of 1.5, $f^{-1}(\{y\}) = \{x\}$ for each y in F and some unique x in E. In this case, we drop some brackets and write $f^{-1}(y) = x$. The resulting function is a bijection from F onto E; it is called the functional inverse of f. This particular usage should not be confused with the general notation of f^{-1}. (Note that 1.5 defines a function f^{-1} from $\mathcal{F}$ into $\mathcal{E}$, where $\mathcal{F}$ is the collection of all subsets of F and $\mathcal{E}$ is the collection of all subsets of E.)

C. Countability

Two sets A and B are said to have the same cardinality if there exists a bijection from A onto B, and then we write $A \sim B$. Obviously, having the same cardinality is an equivalence relation: it is

(a) reflexive, $A \sim A$;
(b) symmetric, $A \sim B \Rightarrow B \sim A$;
(c) transitive, $A \sim B$ and $B \sim C \Rightarrow A \sim C$.

A set is said to be *finite* if it is empty or has the same cardinality as $\{1, 2, \ldots, n\}$ for some n in $\mathbb{N}^*$; in the former case it has 0 elements, in the latter exactly n. It is said to be *countable* if it is finite or has the same cardinality as $\mathbb{N}^*$; in the latter case it is said to have a countable infinity of elements.

In particular, $\mathbb{N}^*$ is countable. So are $\mathbb{Z}$ and $\mathbb{N}^* \times \mathbb{N}^*$ in view of Exercises 1.8 and 1.9. Note that an infinite set can have the same cardinality as one of its proper subsets. For instance, $\mathbb{Z} \sim \mathbb{N}^*$, $\mathbb{R}_+ \sim (0,1]$, $\mathbb{R} \sim \mathbb{R}_+ \sim (0,1)$; see Exercise 1.7 for the latter. Incidentally, $\mathbb{R}_+$, $\mathbb{R}$, etc. are uncountable, as we shall show shortly.

A set is countable if and only if it can be injected into $\mathbb{N}^*$, or equivalently, if and only if there is a surjection from $\mathbb{N}^*$ onto it. Thus, a set A is countable if and only if there is a sequence (x_n) whose range is A. The following lemma follows easily from these remarks.

1.11 LEMMA. If A can be injected into B, and if B is countable, then A is countable. If A is countable and there is a surjection from A onto B, then B is countable.

1.12 THEOREM. The product of two countable sets is countable.

PROOF. Let A and B be countable. If one of them is empty, then $A \times B$ is empty and there is nothing to prove. Suppose that neither is empty. Then, there exist injections $f : A \mapsto \mathbb{N}^*$ and $g : B \mapsto \mathbb{N}^*$. For each pair (x, y) in $A \times B$, let $h(x, y) = (f(x), g(y))$; then h is an injection from $A \times B$ into $\mathbb{N}^* \times \mathbb{N}^*$. Since $\mathbb{N}^* \times \mathbb{N}^*$ is countable (see Exercise 1.9), this implies via the preceding lemma that $A \times B$ is countable. □

1.13 COROLLARY. The set of all rational numbers is countable.

PROOF. Recall that each rational is a ratio m/n with m in $\mathbb{Z}$ and n in $\mathbb{N}^*$. Thus, $f(m, n) = m/n$ defines a surjection from $\mathbb{Z} \times \mathbb{N}^*$ onto the set $\mathbb{Q}$ of all rationals. Since $\mathbb{Z}$ and $\mathbb{N}^*$ are countable, so is $\mathbb{Z} \times \mathbb{N}^*$ by the preceding theorem. Hence, $\mathbb{Q}$ is countable by Lemma 1.11. □

1.14 THEOREM. The union of a countable collection of countable sets is countable.

PROOF. Let I be a countable set, and let A_i be a countable set for each i in I. The claim is that $A = \bigcup_{i \in I} A_i$ is countable. Now, there is a surjection $f_i : \mathbb{N}^* \mapsto A_i$ for each i, and there is a surjection $g : \mathbb{N}^* \mapsto I$; these follow from the countability of I and the A_i. Note that, then, $h(m, n) = f_{g(m)}(n)$ defines a surjection h from $\mathbb{N}^* \times \mathbb{N}^*$ onto A. Since $\mathbb{N}^* \times \mathbb{N}^*$ is countable, this implies via Lemma 1.11 that A is countable. □

The following theorem exhibits an uncountable set. As a corollary, we show that $\mathbb{R}$ is uncountable.

1.15 THEOREM. Let E be the set of all sequences whose terms are the digits 0 and 1. Then, E is uncountable.

PROOF. Let A be a countable subset of E. Let $x_1, x_2, \ldots$ be an enumeration of the elements of A, that is, A is the range of (x_n). Note that each x_n is a sequence of zeros and ones, say $x_n = (x_{n,1}, x_{n,2}, \ldots)$ where each term $x_{n,m}$ is either 0 or 1. We define a new sequence $y = (y_n)$ by letting $y_n = 1 - x_{n,n}$. The sequence y differs from every one of the sequences $x_1, x_2, \ldots$ in at least one position. Thus, y is not in A but is in E.

We have shown that if A is countable, then there is a y in E such that $y \notin A$. If E were countable, the preceding would hold for $A = E$, which would be absurd. Hence, E must be uncountable.   □

1.16 COROLLARY. The set of all real numbers is uncountable.

PROOF. It is enough to show that the interval $[0, 1)$ is uncountable. For each x in $[0, 1)$, let $0.x_1 x_2 x_3 \cdots$ be the binary expansion of x (in case x is dyadic, say $x = k/2^n$ for some k and n in $\mathbb{N}^*$, there are two such possible binary expansions, in which case we take the expansion with infinitely many zeros), and then identify the binary expansion with the sequence $(x_1, x_2, \ldots)$ in the set E of the preceding theorem. Thus, to each x in $[0, 1)$ there corresponds a unique element $f(x)$ of E. In fact, f is a surjection onto the set $E \setminus D$ where D denotes the set of all sequences of zeros and ones that are eventually all ones. It is easy to show that D is countable and hence that $E \setminus D$ is uncountable. It follows that $[0, 1)$ is uncountable.   □

Exercises

1.17 *Dyadics.* A number is said to be dyadic if it has the form $k/2^n$ for some integers k and n in $\mathbb{N}$. Show that the set of all dyadic numbers is countable. Of course, every dyadic number is rational.

1.18 Let D denote the set of all sequences of zeros and ones that are eventually all ones. Show that D is countable.

1.19 Suppose that A is uncountable and that B is countable. Show that $A \setminus B$ is uncountable.

1.20 Let x be a real number. For each n in $\mathbb{N}$, let x_n be the smallest dyadic number of the form $k/2^n$ that exceeds x. Show that $x_0 \geq x_1 \geq x_2 \geq \cdots$ and that $x_n > x$ for each n. Show that, for every $\epsilon > 0$, there is an n_ϵ such that $x_n - x < \epsilon$ for all $n \geq n_\epsilon$.

D. On the Real Line

The object is to review some facts and establish some terminology regarding the set $\mathbb{R}$ of all real numbers and the set $\bar{\mathbb{R}} = [-\infty, +\infty]$ of all extended real numbers. The *extended real number system* consists of $\mathbb{R}$ and two extra symbols, $-\infty$ and ∞. The relation $<$ is extended to $\bar{\mathbb{R}}$ by postulating that $-\infty < x < +\infty$ for every real number x. The arithmetic operations are extended to $\bar{\mathbb{R}}$ as follows: for each x in $\mathbb{R}$,

$$x + \infty = x - (-\infty) = \infty,$$
$$x + (-\infty) = x - \infty = -\infty,$$
$$x \cdot \infty = \begin{cases} \infty & \text{if } x > 0, \\ 0 & \text{if } x = 0, \\ -\infty & \text{if } x < 0, \end{cases}$$
$$x \cdot (-\infty) = (-x) \cdot \infty,$$
$$x/\infty = x/(-\infty) = 0,$$
$$\infty + \infty = \infty,$$
$$(-\infty) + (-\infty) = -\infty,$$
$$\infty \cdot \infty = (-\infty) \cdot (-\infty) = \infty,$$
$$\infty \cdot (-\infty) = -\infty.$$

The operations $(\infty - \infty)$, $(-\infty) - (-\infty)$, $+\infty/+\infty$, $-\infty/-\infty$, and $0/0$ are undefined.

Positive and Negative

We call x in $\bar{\mathbb{R}}$ *positive* if $x \geq 0$ and *strictly positive* if $x > 0$. By symmetry, then, x is *negative* if $x \leq 0$ and strictly negative if $x < 0$. A function $f\colon E \mapsto \bar{\mathbb{R}}$ is said to be *positive* if $f(x) \geq 0$ for all x in E and *strictly positive* if $f(x) > 0$ for all x in E. Negative and strictly negative functions are defined similarly. This usage is in accord with modern tendencies,[1] though at variance with common usage.

Increasing, Decreasing

A function $f\colon \bar{\mathbb{R}} \mapsto \bar{\mathbb{R}}$ is said to be *increasing* if $f(x) \leq f(y)$ whenever $x \leq y$.

---

[1] Often-used concepts should have simpler names. Mind-bending double negatives should be avoided, and as much as possible, mathematical usage should not conflict with ordinary language.

It is said to be *strictly increasing* if $f(x) < f(y)$ whenever $x < y$. Decreasing and strictly decreasing functions are defined similarly.

These notions carry over to functions $f: E \mapsto \bar{\mathbb{R}}$ with $E \subset \mathbb{R}$. In particular, since a sequence is a function on $\mathbb{N}^*$, these notions apply to sequences in $\bar{\mathbb{R}}$. Thus, for example, $(x_n) \subset \bar{\mathbb{R}}$ is increasing if $x_1 \le x_2 \le \cdots$ and is strictly decreasing if $x_1 > x_2 > \cdots$.

Bounds

Let $A \subset \bar{\mathbb{R}}$. A real number b is called an *upper bound* for A provided that $A \subset [-\infty, b]$, and then A is said to be *bounded above* by b. Lower bounds and being bounded below are defined similarly. The set A is said to be *bounded* if it is bounded above and below; that is, if $A \subset [a, b]$ for some real interval $[a, b]$.

These notions carry over to functions and sequences as follows. Given $f: E \mapsto \bar{\mathbb{R}}$, the function f is said to be bounded above, below, etc. accordingly as its range is bounded above, below, etc. Thus, for instance, f is bounded if there exist real numbers $a \le b$ such that $a \le f(x) \le b$ for all x in E.

Supremum and Infimum

These generalize the notions of maximum and minimum. Let $A \subset \bar{\mathbb{R}}$. The *supremum* of A is the smallest number b in $\bar{\mathbb{R}}$ such that $A \subset [-\infty, b]$; the *infimum* of A is the largest number a in $\bar{\mathbb{R}}$ such that $A \subset [a, \infty]$. They are denoted, respectively,

$$\sup A, \quad \inf A.$$

In particular, $\sup \emptyset = -\infty$, $\inf \emptyset = +\infty$, $\inf(a, b] = \inf[a, b] = a$, and $\sup(a, b) = \sup(a, b] = b$. For $A = \{1, 1/2, 1/3, \dots\}$, the supremum is 1 and the infimum is 0.

If A is finite, then $\inf A$ is the smallest element of A, and $\sup A$ is the largest. Even when A is infinite, it is possible that $\inf A$ is an element of A, in which case it is called the *minimum* of A. Similarly, if $\sup A$ is an element of A, then it is also called the *maximum* of A.

If $f: E \mapsto \bar{\mathbb{R}}$ and $D \subset E$, it is customary to write

$$\inf_{x \in D} f(x) = \inf\{f(x) : x \in D\}$$

and call it the infimum of f over D, and similarly with the supremum. In the case of sequences (x_n) in $\bar{\mathbb{R}}$,

$$\inf x_n, \quad \sup x_n$$

denote, respectively, the infimum and supremum of the range of (x_n). Other such notations are generally self-explanatory; for example,

$$\inf_{n \ge k} x_n = \inf\{x_k, x_{k+1}, \dots\}, \quad \sup_{k \ge 1} x_{nk} = \sup\{x_{n1}, x_{n2}, \dots\}.$$

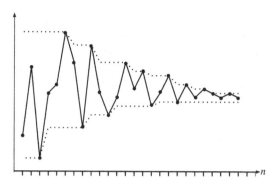

FIGURE 1.1. lim sup and lim inf. The pairs (n, x_n) are connected by the *solid lines* for clarity. The pairs $(n, \underline{x}_n)$ form the *lower dotted line* and $(n, \bar{x}_n)$ the *upper dotted line*.

Limits

If (x_n) is an increasing sequence in $\bar{\mathbb{R}}$, then $\sup x_n$ is also called the *limit* of (x_n) and is denoted by $\lim x_n$. If it is a decreasing sequence, then $\inf x_n$ is called the limit of (x_n) and again denoted by $\lim x_n$.

Let (x_n) be an arbitrary sequence in $\bar{\mathbb{R}}$. Then

1.21
$$\underline{x}_m = \inf_{n \geq m} x_n, \quad \bar{x}_m = \sup_{n \geq m} x_n, \quad m \in \mathbb{N}^*,$$

define two sequences; $(\underline{x}_n)$ is increasing, and $(\bar{x}_n)$ is decreasing. Their limits are called the *limit inferior* and the *limit superior*, respectively, of the sequence (x_n):

1.22
$$\liminf x_n = \lim \underline{x}_n = \sup_m \inf_{n \geq m} x_n,$$

1.23
$$\limsup x_n = \lim \bar{x}_n = \inf_m \sup_{n \geq m} x_n,$$

Figure 1.1 is worthy of careful study. Note that, in general,

1.24
$$-\infty \leq \liminf x_n \leq \limsup x_n \leq +\infty.$$

If $\liminf x_n = \limsup x_n$, then the common value is called the *limit* of (x_n) and is denoted by $\lim x_n$. Otherwise, if limits inferior and superior are not equal, the sequence (x_n) does not have a limit.

Convergence of Sequences

A sequence (x_n) of real numbers is said to be *convergent* if $\lim x_n$ exists and is a real number.

An examination of Fig. 1.1 shows that this is equivalent to the classical definition of convergence: (x_n) converges to x if for every $\epsilon > 0$ there is an integer n_ϵ such that $|x_n - x| < \epsilon$ for all $n \geq n_\epsilon$. The phrase "there is $n_\epsilon \ldots$ for all $n \geq n_\epsilon$" can be expressed in more geometric terms by phrases like "the number of terms outside $(x - \epsilon, x + \epsilon)$ is finite," or "all but finitely many terms are in $(x - \epsilon, x + \epsilon)$," or "$|x_n - x| < \epsilon$ for all n large enough."

The following is a summary of the relations between convergence and algebraic operations. The proof will be omitted.

1.25 THEOREM. Let (x_n) and (y_n) be convergent sequences with limits x and y respectively. Then,

    (a) $\lim cx_n = cx$,
    (b) $\lim(x_n + y_n) = x + y$,
    (c) $\lim x_n y_n = xy$,
    (d) $\lim x_n/y_n = x/y$ provided that $y_n, y \neq 0$.

In practice, we do not have the sequence laid out before us. Instead, some rule is given for generating the sequence and the object is to show that the resulting sequence will converge. For instance, a function may be specified somehow and a procedure described to find its maximum; starting from some point, the procedure will give the successive points $x_1, x_2, \ldots$ which are meant to form the sequence that converges to the point x where the maximum is achieved.

Often, to find the limit of (x_n), one starts with a search for sequences that bound (x_n) from above and below and whose limits can be computed easily: supposing that

$$y_n \leq x_n \leq z_n \quad \text{for all } n, \qquad \lim y_n = \lim z_n,$$

then $\lim x_n$ exists and is equal to the limit of the other two. The art involved is in finding such sequences (y_n) and (z_n).

1.26 EXAMPLE. This is to illustrate the technique mentioned above. We want to show that $(n^{1/n})$ converges. Note that $n^{1/n} \geq 1$ always, so put $x_n = n^{1/n} - 1$, and consider the sequence (x_n). Now, $(1 + x_n)^n = n$, and by the binomial theorem

$$(a + b)^n = a^n + na^{n-1}b + \frac{n(n-1)}{2}a^{n-2}b^2 + \cdots + b^n \geq \frac{n(n-1)}{2}a^{n-2}b^2$$

for $a, b \geq 0$ and $n \geq 2$. So,

$$n = (1 + x_n)^n \geq \frac{n(n-1)}{2}x_n^2,$$

or

$$0 \leq x_n \leq \sqrt{2/(n-1)}.$$

It follows that $\lim x_n = 0$, and hence

$$\lim n^{1/n} = 1.$$

Exercises

1.27 Show that if $A \supset B \neq \emptyset$ then $\inf A \leq \inf B \leq \sup B \leq \sup A$. Use this to show that, if $A_1 \supset A_2 \supset \cdots$, then

$$\inf A_1 \leq \inf A_2 \leq \cdots \leq \inf A_n \leq \cdots \leq \sup A_n \leq \cdots \leq \sup A_2 \leq \sup A_1.$$

Use this to show that $(\underline{x}_n)$ is increasing, $(\bar{x}_n)$ is decreasing, and $\lim \underline{x}_n \leq \lim \bar{x}_n$ (see 1.21–1.23 for definitions).

1.28 Show that $\sup(-x_n) = -\inf x_n$ for any sequence (x_n) in $\bar{\mathbb{R}}$. Conclude that $\limsup(-x_n) = -\liminf x_n$.

1.29 *Cauchy criterion.* Sequence (x_n) is convergent if and only if for every $\epsilon > 0$ there is an integer n_ϵ such that $|x_m - x_n| \leq \epsilon$ for all $m \geq n \geq n_\epsilon$. Prove this by examining Fig. 1.1 on the definition of the limit.

1.30 *Monotone sequences.* If (x_n) is increasing, then $\lim x_n$ exists (but could be $+\infty$). Thus, such a sequence converges if and only if it is bounded above. Show this. State the version of this for decreasing sequences.

1.31 *Iterative sequences.* Often, x_{n+1} is obtained from x_n via some rule, that is, $x_{n+1} = f(x_n)$ for some function f, in which case the sequence is said to be iterative. If (x_n) is such a sequence and f is continuous and $\lim x_n = x$ exists, then $x = f(x)$. This works well for identifying the limit especially when f is simple and $x = f(x)$ has only one solution. In general, with complicated functions f, the reverse is true: To find x satisfying $x = f(x)$, one starts at some point x_0, computes $x_1 = f(x_0)$, $x_2 = f(x_1)$, ... and tries to show that $x = \lim x_n$ exists and satisfies $x = f(x)$.

1.32 *Domination.* A sequence (x_n) is said to be dominated by a sequence (y_n) if $x_n \leq y_n$ for each n. Show that, if this is the case, then
   (a) $\inf x_n \leq \inf y_n$,
   (b) $\sup x_n \leq \sup y_n$,
   (c) $\liminf x_n \leq \liminf y_n$,
   (d) $\limsup x_n \leq \limsup y_n$.
In particular, if the limits exist, $\lim x_n \leq \lim y_n$.

Incidentally, $(\underline{x}_n)$ defined by 1.21 is the maximal increasing sequence dominated by (x_n), and $(\bar{x}_n)$ is the minimal decreasing sequence dominating (x_n).

1.33 *Comparisons.* Let (x_n) be a positive sequence. Then, (x_n) converges to 0 if and only if it is dominated by a sequence (y_n) with $\limsup y_n = 0$. Show this.

Favorite sequences (y_n) used in this role are given by $y_n = 1/n$, $y_n = r^n$ for some fixed number $r \in (0,1)$, and $y_n = n^p r^n$ with $p \in (-\infty, +\infty)$ and $r \in (0,1)$.

1.34 *Existence of least upper bounds.* Let A be a nonempty subset of $\mathbb{R}$ and let $B = \{b : b \text{ is an upper bound of } A\}$. Assuming that B is nonempty, show that B is an interval of the form $B = [a, \infty]$. Thus a is the minimum of B and is the definition of $\sup A$.

E. Series

Given a sequence (x_n) in $\mathbb{R}$, the sequence (s_n) defined by

1.35
$$s_n = \sum_{i=1}^{n} x_i$$

is called the sequence of partial sums of (x_n), and the symbolic expression

1.36
$$\sum x_n$$

is called the *series* associated with (x_n). The series is said to *converge* to s if and only if the sequence (s_n) converges to s, and then we write

1.37
$$\sum_{1}^{\infty} x_n = s.$$

Sometimes, we write $x_1 + x_2 + \cdots$ for the series 1.36. Sometimes, for convenience of notation, we shall consider series of the form $\sum_0^\infty$ or $\sum_m^\infty$, depending on the index set. Here are a few examples:

$$\sum_{n=0}^{\infty} x^n = \frac{1}{1-x}, \quad x \in (-1,1),$$

$$\sum_{n=0}^{\infty} \frac{x^n}{n!} = e^x, \quad x \in \mathbb{R},$$

$$\sum_{n=1}^{\infty} \frac{1}{n^2} = \frac{\pi^2}{6},$$

$$\sum_{n=m}^{\infty} x^n = \frac{x^m}{1-x}, \quad x \in (-1,1).$$

The following result is obtained by applying the Cauchy criterion (Exercise 1.29) to the sequence of partial sums.

1.38 THEOREM. The series $\sum x_n$ converges if and only if for every $\epsilon > 0$ there is an integer n_ϵ such that

1.39
$$\left| \sum_{i=n}^{m} x_i \right| \leq \epsilon$$

for all $m \geq n \geq n_\epsilon$.

In particular, taking $m = n$ in 1.39 we obtain $|x_n| \leq \epsilon$. Thus we have the following consequence.

1.40 COROLLARY. If $\sum x_n$ converges, then $\lim x_n = 0$.

The converse is not true. For example, $\lim 1/n = 0$ but $\sum 1/n$ is divergent. In the case of series with positive terms, partial sums form an increasing sequence, and hence, the following proposition holds (see Exercise 1.30).

1.41 PROPOSITION. Suppose that the terms x_n are positive. Then $\sum x_n$ converges if and only if the sequence of partial sums is bounded.

In many cases, we encounter series whose terms are positive and decreasing. The following theorem due to Cauchy is helpful in such cases, especially if the terms involve powers. Note the way a rather thin sequence determines the convergence or divergence of the whole series.

1.42 THEOREM. Suppose that (x_n) is decreasing and positive. Then $\sum x_n$ converges if and only if the series
$$x_1 + 2x_2 + 4x_4 + 8x_8 + \cdots$$
converges.

PROOF. Let $s_n = x_1 + \cdots + x_n$ as usual and put $t_k = x_1 + 2x_2 + \cdots + 2^k x_{2^k}$. Now, for $n \leq 2^k$, since $x_1 \geq x_2 \geq \cdots \geq 0$,

$$
\begin{aligned}
s_n &\leq x_1 + (x_2 + x_3) + (x_4 + \cdots + x_7) + \cdots + (x_{2^k} + \cdots + x_{2^{k+1}-1}) \\
&\leq x_1 + 2x_2 + 4x_4 + \cdots + 2^k x_{2^k} \\
&= t_k,
\end{aligned}
$$

and for $n \geq 2^k$,

$$
\begin{aligned}
s_n &\geq x_1 + x_2 + (x_3 + x_4) + (x_5 + \cdots + x_8) + \cdots + (x_{2^{k-1}+1} + \cdots + x_{2^k}) \\
&\geq \frac{1}{2}x_1 + x_2 + 2x_4 + \cdots + 2^{k-1}x_{2^k} \\
&= \frac{1}{2}t_k.
\end{aligned}
$$

Thus, the sequences (s_n) and (t_n) are either both bounded or both unbounded, which completes the proof via Proposition 1.41. □

1.43 EXAMPLE. $\sum 1/n^p$ converges if $p > 1$ and diverges if $p \leq 1$. For $p \leq 0$, the claim is trivial to see. For $p > 0$, the terms $x_n = 1/n^p$ form a decreasing positive sequence, and thus, the preceding theorem applies. Now, writing $x(n)$ for x_n,

$$
\sum_{k=0}^{\infty} 2^k x(2^k) = \sum (2^{1-p})^k,
$$

which converges if $2^{1-p} < 1$ and diverges otherwise. Since $2^{1-p} < 1$ if and only if $p > 1$, we are done.

1.44 EXAMPLE. The series

$$
\sum_{2}^{\infty} \frac{1}{n(\log n)^p}
$$

converges if $p \in (1, \infty)$ and diverges otherwise. Here we start the series with $n = 2$ since $\log 1 = 0$. Since the logarithm function is monotone increasing, Theorem 1.42 applies. Now, $x(n) = 1/n(\log n)^p$, and so

$$
\sum_{k=1}^{\infty} 2^k x(2^k) = \sum_{1}^{\infty} 2^k \frac{1}{2^k (\log 2^k)^p} = \frac{1}{(\log 2)^p} \sum_{1}^{\infty} \frac{1}{k^p},
$$

which converges if and only if $p > 1$ in view of the preceding example.

Ratio Test, Root Test

The ratio test ties the convergence of $\sum x_n$ to the behavior of the ratios x_{n+1}/x_n for large n; it is highly useful.

1.45 THEOREM. We have the following:

(a) If $\limsup |x_{n+1}/x_n| < 1$, then $\sum x_n$ converges.
(b) If $\liminf |x_{n+1}/x_n| > 1$, then $\sum x_n$ diverges.

PROOF. If $\limsup |x_{n+1}/x_n| < 1$, then there is a number r in $[0,1)$ and an integer n_0 such that $|x_{n+1}/x_n| \le r$ for all $n \ge n_0$. Thus $|x_{n_0+k}| \le |x_{n_0}| r^k$ for all $k \ge 0$, and therefore, for $m > n > n_0$,

$$\left| \sum_{i=n}^{m} x_i \right| \le \sum_{i=n}^{\infty} |x_i| \le |x_{n_0}| \sum_{i=n}^{\infty} r^{i-n_0} = |x_{n_0}| \frac{r^{n-n_0}}{1-r}.$$

Given $\epsilon > 0$ choose n_ϵ so that $|x_{n_0}| r^{n_\epsilon - n_0}/(1-r) < \epsilon$. Then Cauchy's criterion works with this n_ϵ and $\sum x_n$ converges.

If $\liminf |x_{n+1}/x_n| > 1$ then there is an integer n_0 such that $|x_{n+1}| \ge |x_n|$ for all $n \ge n_0$. Hence, $|x_n| \ge |x_{n_0}|$ for all $n \ge n_0$ which shows that (x_n) does not converge to 0 as it must in order for $\sum x_n$ to converge (see Corollary 1.40).   □

The preceding test gives no information in cases where

$$\liminf |x_{n+1}/x_n| \le 1 \le \limsup |x_{n+1}/x_n|.$$

For instance, for the two series $\sum 1/n$ and $\sum 1/n^2$, the preceding lim inf and lim sup are equal to 1, but the first series diverges whereas the second converges. Also, the series

1.46
$$\frac{1}{2} + \frac{1}{3} + \frac{1}{2^2} + \frac{1}{3^2} + \frac{1}{2^3} + \frac{1}{3^3} + \frac{1}{2^4} + \frac{1}{3^4} + \cdots .$$

obviously converges to $3/2$; yet, the ratio test is miserably inconclusive:

$$\liminf \frac{x_{n+1}}{x_n} = \lim \left(\frac{2}{3} \right)^n = 0,$$

$$\limsup \frac{x_{n+1}}{x_n} = \lim \left(\frac{3}{2} \right)^n = \infty.$$

The following test, called the *root test*, is a stronger test—if the ratio test works, so does the root test. But the root test works in some situations where the ratio test fails; for example, the root test works for the series 1.46.

1.47 THEOREM. Let $a = \limsup |x_n|^{1/n}$. Then $\sum x_n$ converges if $a < 1$, and diverges if $a > 1$.

PROOF. Suppose that $a < 1$. Then, there is a number $b \in (a,1)$ such that $|x_n|^{1/n} \le b$ for all $n \ge n_0$, where n_0 is some integer. Then, $|x_n| \le b^n$ for all $n \ge n_0$, and comparing $\sum x_n$ with the geometric series $\sum b^n$ shows that $\sum x_n$ converges.

Suppose that $a > 1$. Then, a subsequence of $|x_n|$ must converge to $a > 1$, which means that $|x_n| \geq 1$ for infinitely many n. So, (x_n) does not converge to zero, and hence, $\sum x_n$ cannot converge.   □

Power Series

The convergence of series is extended to series of complex numbers as follows. Each complex number z can be thought of as a pair (x, y) of real numbers, or better yet as $z = x + iy$ where $i = \sqrt{-1}$. For a sequence (z_n) of complex numbers, the sequence is said to converge to z if (x_n) converges to x and (y_n) converges to y, where $z_n = x_n + iy_n$ and $z = x + iy$.

Given a sequence (c_n) of complex numbers, the series

1.48
$$\sum_{0}^{\infty} c_n z^n$$

is called a *power series*. The numbers $c_0, c_1, \ldots$ are called the coefficients of the power series; here z is a complex number.

In general, the series will converge or diverge depending on the choice of z. As the following theorem shows, there is a number $r \in [0, \infty]$, called the radius of convergence, such that the series converges if $|z| < r$ and diverges if $|z| > r$. The behavior for $|z| = r$ is much more complicated and cannot be described easily.

1.49 THEOREM. Let $a = \limsup |c_n|^{1/n}$ and $r = 1/a$.

  (a) If $|z| < r$, then $\sum c_n z^n$ converges.
  (b) If $|z| > r$, then $\sum c_n z^n$ diverges.

PROOF. Put $x_n = c_n z^n$ and apply the root test with

$$\limsup |x_n|^{1/n} = |z| \limsup |c_n|^{1/n} = a|z| = \frac{|z|}{r}.$$

                                                                                        □

1.50 EXAMPLE.

  (a) $\sum z^n/n! = e^z$ and $r = \infty$.
  (b) $\sum z^n$ converges for $|z| < 1$ and diverges for $|z| \geq 1$; $r = 1$.
  (c) $\sum z^n/n^2$ converges for $|z| \leq 1$ and diverges for $|z| > 1$; $r = 1$.

(d) $\sum z^n/n$ converges for $|z| < 1$ and diverges for $|z| > 1$; $r = 1$; for $z = 1$ the series diverges, but for $|z| = 1$ with $z \neq 1$ it converges.

Absolute Convergence

    A series $\sum x_n$ is said to *converge absolutely* if $\sum |x_n|$ is convergent. If the x_n are all positive numbers, then absolute convergence is the same as convergence. Using Cauchy's criterion (see Theorem 1.38) on both sides of

$$\left| \sum_{i=n}^{m} x_i \right| \leq \sum_{i=n}^{m} |x_i|$$

shows that if $\sum x_n$ converges absolutely then it converges. But the converse is not true: for example,

$$\sum (-1)^n/n$$

converges but is not absolutely convergent.

    The comparison tests above, as well as the root and ratio tests, are in fact tests for absolute convergence. If a series is not absolutely convergent, one has to study the sequence of partial sums to determine whether the series converges at all.

Rearrangements

    Let $(k_1, k_2, \dots)$ be a sequence in which every integer $n \geq 1$ appears once and only once, that is, $n \mapsto k_n$ is a bijection from $\mathbb{N}^*$ onto $\mathbb{N}^*$. If

$$y_n = x_{k_n}, \quad n \in \mathbb{N}^*,$$

for such a sequence (k_n), then we say that (y_n) is a rearrangement of (x_n).

    Let (y_n) be a rearrangement of (x_n). In general, the series $\sum y_n$ and $\sum x_n$ are quite different. However, if $\sum x_n$ is absolutely convergent, then so is $\sum y_n$ and it converges to the same number as does $\sum x_n$. The converse is also true: if every rearrangement of the series $\sum x_n$ converges, then the series $\sum x_n$ is absolutely convergent and all its rearrangements converge (to the same sum).

    On the other hand, if $\sum x_n$ is not absolutely convergent, its various rearrangements may converge or diverge, and in the case of convergence, the sum generally depends on the rearrangement chosen. For instance,

$$1 - \frac{1}{2} + \frac{1}{3} - \cdot\frac{1}{4} + \frac{1}{5} - \frac{1}{6} + \frac{1}{7} - \cdots$$

is convergent, but not absolutely so. Its rearrangement

$$1 + \frac{1}{3} - \frac{1}{2} + \frac{1}{5} + \frac{1}{7} - \frac{1}{4} + \frac{1}{9} + \cdots$$

(with $++-++-++-$ pattern) is again convergent, but not to the same sum. In fact, the following theorem due to Riemann shows that one can create rearrangements that are as bizarre as one wants.

1.51 THEOREM. Let $\sum x_n$ be convergent but not absolutely. Then, for any two numbers $a \le b$ in $\bar{\mathbb{R}}$ there is a rearrangement $\sum y_n$ of $\sum x_n$ such that

$$\liminf \sum_1^n y_i = a, \quad \limsup \sum_1^n y_i = b.$$

We omit the proof. Note that, in particular, taking $a = b$ we can find a rearrangement $\sum y_n$ with sum a, no matter what a is.

Exercises

1.52   Determine the convergence or divergence of the following:

(a) $\sum(\sqrt{n+1} - \sqrt{n})$
(b) $\sum(\sqrt{n+1} - \sqrt{n})/n$
(c) $\sum(\sin n)/(n\sqrt{n})$
(d) $\sum(-1)^n n/(n^2 + 1)$

In case of convergence, indicate whether it is absolute convergence.

1.53   Show that if $\sum x_n$ converges then so does $\sum \sqrt{x_n}/n$.

1.54   Show that if $\sum x_n$ converges and (y_n) is bounded and monotone (either increasing or decreasing), then $\sum x_n y_n$ converges.

1.55   Find the radius of convergence of each of the following power series:

(a) $\sum n^2 z^n$
(b) $\sum 2^n z^n /n!$
(c) $\sum 2^n z^n /n^2$
(d) $\sum n^3 z^n /3^n$

1.56   Suppose that $f(z) = \sum c_n z^n$. Express the sum of the even terms, $\sum c_{2n} z^{2n}$, and the sum of the odd terms, $\sum c_{2n+1} z^{2n+1}$, in terms of f.

1.57   Suppose that $f(z) = \sum c_n z^n$. Express $\sum c_{3n} z^{3n}$ in terms of f.

1.58 *Rearrangements.* Let $\sum x_n$ be a series that converges absolutely. Prove that every rearrangement of $\sum x_n$ converges, and that they all converge to the same sum.

1.59 *Riemann's theorem.* Prove Riemann's theorem 1.51 by filling in the details in the following outline:

(a) Let (x_n^+) denote the subsequence consisting of the positive elements of (x_n) and let (x_n^-) denote the subsequence of negative elements of (x_n). Both of these sequences must be infinite.

(b) Both sequences (x_n^+) and (x_n^-) converge to zero.

(c) Both series $\sum x_n^+$ and $\sum x_n^-$ diverge.

(d) Suppose that $a, b \in \mathbb{R}$ and define a rearrangement as follows: start with the positive elements and choose elements from this set until the partial sum exceeds b. Then, choose elements from the set of negative elements until the partial sum is less than a. Then, choose elements from the set of positive elements until the partial sum exceeds b. Continue this procedure of alternating between elements of the positive and negative sets indefinitely.

(e) Prove that the procedure described above can be continued ad infinitum.

(f) Prove that this rearrangement has the properties stated in Riemann's theorem.

(g) Extend the above arguments to the case where $a, b = \pm\infty$.

1.60 *Poisson distribution.* Let $p_n = e^{-\lambda}\lambda^n/n!$ where λ is a strictly positive real number. Show that

(a) $p_n > 0$,

(b) $\sum_{n=0}^{\infty} p_n = 1$,

(c) $\sum_{n=0}^{\infty} np_n = \lambda$.

1.61 *Borel summability.* Consider a series $\sum_{n=0}^{\infty} x_n$ with partial sums $s_n = \sum_{i=0}^{n} x_i$. We say that the series is *Borel summable* if

$$\lim_{\lambda \to \infty} \sum_{n=0}^{\infty} s_n p_n$$

converges, where p_n are the Poisson probabilities defined in Exercise 1.60. For what values of z is the geometric series $\sum_{n=0}^{\infty} z^n$ Borel summable?

CHAPTER 2

Metric Spaces

Basic questions of analysis on the real line are tied to the notions of closeness and distances between points. The same issue of closeness comes up in more complicated settings, for instance, when we try to approximate a function by a simpler function. Our aim is to introduce the idea of distance in general so that we can talk of the distance between two functions with the same conceptual ease as when we talk of the distance between two points in a plane. After that, we discuss the main issues: convergence, continuity, approximations. All along, there will be examples of different spaces and different ways of measuring distances.

A. Euclidean Spaces

This section will review the space $\mathbb{R}^n$ together with its Euclidean distance. Recall that each element of $\mathbb{R}^n$ is an n-tuple $x = (x_1, \ldots, x_n)$, where the x_i are real numbers. The elements of $\mathbb{R}^n$ are called *points* or *vectors*, and we are familiar with vector addition and scalar multiplication.

Inner Product and Norm

For x and y in $\mathbb{R}^n$, their *inner product* $x \cdot y$ is the number

2.1
$$x \cdot y = \sum_1^n x_i y_i.$$

If we regard x and y as column vectors, then $x \cdot y = x^T y$, where x^T is the transpose of x, the row vector with the same elements as x. For x in $\mathbb{R}^n$, the *norm* of x is defined as the length of x, that is, the positive number

2.2
$$\|x\| = \sqrt{x \cdot x} = \sqrt{\sum_1^n x_i^2}.$$

The norm satisfies the following conditions:

2.3
$$\|x\| \geq 0 \text{ for every } x \text{ in } \mathbb{R}^n,$$

2.4
$$\|x\| = 0 \text{ if and only if } x = 0,$$

E. Çınlar and R.J. Vanderbei, *Real and Convex Analysis*, Undergraduate Texts in Mathematics, DOI 10.1007/978-1-4614-5257-7_2, © Springer Science+Business Media New York 2013

2.5 $\|\lambda x\| = |\lambda|\|x\|$ for every λ in $\mathbb{R}$,

2.6 $\|x + y\| \leq \|x\| + \|y\|$ for all x and y in $\mathbb{R}^n$.

Of these, 2.3–2.5 are obvious, and 2.6 is an immediate consequence of the following proposition.

2.7 PROPOSITION. *Schwarz's inequality.* $|x \cdot y| \leq \|x\|\|y\|$ for all x and y in $\mathbb{R}^n$.

PROOF. Fix x and y. Consider the function f with

$$f(\lambda) = \|\lambda y - x\|^2 = \lambda^2 \|y\|^2 - 2\lambda(x \cdot y) + \|x\|^2.$$

This function is clearly positive and quadratic, and its minimum occurs at

$$\lambda = \frac{x \cdot y}{\|y\|^2}.$$

For this value of λ we have

$$0 \leq f(\lambda) = -\frac{(x \cdot y)^2}{\|y\|^2} + \|x\|^2,$$

from which the claimed inequality follows.   □

Euclidean Distance

For x and y in $\mathbb{R}^n$, the *Euclidean distance* between x and y is defined as the number $\|x - y\|$. It follows from the properties given above that, for all x, y, z in $\mathbb{R}^n$,

2.8 $\|x - y\| \geq 0,$

2.9 $\|x - y\| = \|y - x\|,$

2.10 $\|x - y\| = 0$ if and only if $x = y,$

2.11 $\|x - y\| + \|y - z\| \geq \|x - z\|.$

The last is called the *triangle inequality*: on $\mathbb{R}^2$, if the points x, y, z are the vertices of a triangle, this is simply the well-known fact that the sum of the lengths of two sides is greater than or equal to the length of the third side.

The set $\mathbb{R}^n$ together with the Euclidean distance is called *n-dimensional Euclidean space*. The Euclidean spaces are important examples of metric spaces.

Exercises

2.12 *Inner product.* Show that the mapping $(x, y) \mapsto x \cdot y$ from $\mathbb{R}^n \times \mathbb{R}^n$ into $\mathbb{R}$ is a linear transformation in x and is a linear transformation in y (and therefore is said to be bilinear). Conclude that

$$(x + y) \cdot (x + y) = x \cdot x + 2x \cdot y + y \cdot y.$$

Use this and the Schwarz inequality to prove 2.6.

2.13 *Parallelograms.* Show that $\|x + y\|^2 + \|x - y\|^2 = 2\|x\|^2 + 2\|y\|^2$. Interpret this in geometric terms, on $\mathbb{R}^2$, as a statement about parallelograms.

2.14 *Orthogonality.* Points x and y are said to be *orthogonal* if $x \cdot y = 0$. Show that this is equivalent to saying that the line segments connecting the origin to x and y are perpendicular. In general, letting α be the angle between these line segments, we have $x \cdot y = \|x\|\|y\| \cos \alpha$.

B. Metrics

Let E be a nonempty set. A *metric* on E is a function $d \colon E \times E \mapsto \mathbb{R}_+$ that satisfies the following for all x, y, z in E:

    (a) $d(x, y) = d(y, x)$,
    (b) $d(x, y) = 0$ if and only if $x = y$,
    (c) $d(x, y) + d(y, z) \geq d(x, z)$.

A *metric space* is a pair (E, d) where E is a set and d is a metric on E. In this context, we think of E as a space, call the elements of E points, and refer to $d(x, y)$ as the distance from x to y.

Examples

2.15 *Euclidean spaces.* Consider $\mathbb{R}^n$ with the Euclidean distance $d(x, y) = \|x - y\|$ on it. It follows from 2.8 to 2.11 that d is a metric on $\mathbb{R}^n$. Thus, $(\mathbb{R}^n, d)$ is a metric space; it is called the n-dimensional Euclidean space.

2.16 *Manhattan metric.* On $\mathbb{R}^n$ define a metric d by

$$d(x,y) = \sum_1^n |x_i - y_i|.$$

This d is called the Manhattan metric, or l_1-metric, on $\mathbb{R}^n$, and $(\mathbb{R}^n, d)$ is a metric space again. Note that for $n > 1$ this metric is different from the Euclidean metric of the preceding example.

2.17 *Space $\mathcal{C}$.* Let $\mathcal{C}$ denote the set of all continuous functions from the interval $[0, 1]$ into $\mathbb{R}$. For x and y in $\mathcal{C}$, let

$$d(x,y) = \sup_{0 \le t \le 1} |x(t) - y(t)|.$$

It is clear that $d(x,y)$ is a positive real number, that $d(x,y) = d(y,x)$, and that $d(x,y) = 0$ if and only if $x = y$. As for the triangle inequality, we note that

$$|x(t) - z(t)| \le |x(t) - y(t)| + |y(t) - z(t)| \le d(x,y) + d(y,z)$$

for every t in $[0, 1]$, from which we have $d(x,z) \le d(x,y) + d(y,z)$. Thus, d is a metric on $\mathcal{C}$, and $(\mathcal{C}, d)$ is a metric space. This metric space is important in analysis.

Usage

In the literature, it is common practice to call E a metric space if (E, d) is a metric space for some metric d. If there is only one metric under consideration, this is harmless and saves time. For instance, the phrase "Euclidean space $\mathbb{R}^n$" refers to $(\mathbb{R}^n, d)$ where d is the Euclidean metric. For a while at least, we shall indicate the metric involved in each case in order to avoid all possible confusion.

Distances from Points to Sets and from Sets to Sets

Let (E, d) be a metric space. For x in E and A a subset of E, let

2.18 $d(x, A) = \inf\{d(x,y) : y \in A\};$

this is called the distance from the point x to the set A. For subsets A and B of E, the distance from A to B is defined by

2.19 $d(A, B) = \inf\{d(x,y) : x \in A, y \in B\}.$

The *diameter* of a set $A \subset E$ is defined as

2.20 $\operatorname{diam} A = \sup\{d(x,y) : x \in A, y \in A\}.$

A set is said to be *bounded* if its diameter is finite.

Balls

Let (E, d) be a metric space. For x in E and r in $(0, \infty)$,

2.21
$$B(x, r) = \{y \in E : d(x, y) < r\}$$

is called the *open ball* with *center* x and *radius* r, and

2.22
$$\bar{B}(x, r) = \{y \in E : d(x, y) \le r\}$$

is the corresponding *closed ball*.

For example, if $E = \mathbb{R}^3$ and d is the usual Euclidean metric, then $B(x, r)$ becomes the set of all points inside the sphere with center x and radius r, and $\bar{B}(x, r)$ is the set of all points inside or on that sphere.

Exercises and Complements

2.23 *Discrete metric.* Let E be an arbitrary nonempty set. Define

$$d(x, y) = \begin{cases} 1 & \text{if } x \ne y, \\ 0 & \text{if } x = y. \end{cases}$$

Show that this d is a metric on E. It is called the discrete metric on E.

2.24 *Metrics on $\mathbb{R}^n$.* For each number $p \ge 1$,

$$d_p(x, y) = \left(\sum_1^n |x_i - y_i|^p \right)^{1/p}$$

defines a metric d_p on $\mathbb{R}^n$. Note that d_1 is the Manhattan metric, and d_2 is the Euclidean metric. Finally,

$$d_\infty(x, y) = \sup_{1 \le i \le n} |x_i - y_i|$$

is again a metric on $\mathbb{R}^n$. Show this last statement.

2.25 *Equivalent metrics.* Two metrics d and d' are equivalent if there exist strictly positive constants c_1 and c_2 such that for all x, y

$$c_1 d'(x, y) \le d(x, y) \le c_2 d'(x, y).$$

Show that d_1, d_2, and d_∞ are all equivalent to each other as distances on $\mathbb{R}^n$.

2.26 *Weighted metrics on* $\mathbb{R}^n$. The metrics introduced in Exercise 2.24 treat all components of x and y equally. This is reasonable if $\mathbb{R}^n$ is thought of geometrically and the selection of a coordinate system is unimportant. On the other hand, if $x = (x_1, \ldots, x_n)$ stands for a shopping list that requires buying x_1 units of product one, and x_2 units of product two, and so on, then it would be better to define the distance between two shopping lists x and y by

$$d(x, y) = \sum_1^n w_i |x_i - y_i|$$

where $w_1, \ldots, w_n$ are fixed strictly positive numbers, with w_i being the value of one unit of product i. Show that this d is indeed a metric. More generally, paralleling the metrics introduced in Exercise 2.24,

$$d_p(x, y) = \left(\sum_i^n w_i |x_i - y_i|^p \right)^{1/p}, \quad x, y \in \mathbb{R}^n,$$

is a metric on $\mathbb{R}^n$ for each $p \geq 1$ and each fixed strictly positive vector w (the latter means $w_1 > 0, \ldots, w_n > 0$).

2.27 l^2-*Spaces.* Instead of $\mathbb{R}^n$, now consider the space $\mathbb{R}^\infty$ of all infinite sequences in $\mathbb{R}$, that is, each x in $\mathbb{R}^\infty$ is a sequence $x = (x_1, x_2, \ldots)$ of real numbers. In analogy with the d_2 metrics introduced on $\mathbb{R}^n$ in Exercises 2.24 and 2.26, we define

$$d_2(x, y) = \left(\sum_1^\infty |x_i - y_i|^2 \right)^{1/2}.$$

This d_2 satisfies all the conditions for a metric except that $d_2(x, y)$ can be ∞ for some x and y. To remedy the latter, we let E be the set of all x in $\mathbb{R}^\infty$ with

$$\sum_1^\infty x_i^2 < \infty.$$

Then, by an easy generalization of the Schwarz inequality, it follows that $d_2(x, y) < \infty$ for all x and y in E. Thus, (E, d_2) is a metric space. It is generally denoted by l^2.

2.28 *Metrics on* $\mathcal{C}$. Consider the set $\mathcal{C}$ of all continuous functions from $[0, 1]$ into $\mathbb{R}$. The interval $[0, 1]$ can be replaced by any bounded interval $[a, b]$, in which case one writes $\mathcal{C}([a, b])$. A number of metrics can be defined on $\mathcal{C}$ in analogy with those in Exercise 2.24 by observing that every x in $\mathbb{R}^n$ can be thought of as a function x from $\{\frac{1}{n}, \frac{2}{n}, \ldots, \frac{n}{n}\}$ into $\mathbb{R}$, namely, the function x with $x(t) = x_i$ for $t = i/n$. Thus, replacing the set $\{\frac{1}{n}, \frac{2}{n}, \ldots, \frac{n}{n}\}$ with the interval $[0, 1]$ and replacing the summation by integration, we obtain

$$d_p(x, y) = \left(\int_0^1 |x(t) - y(t)|^p \, dt \right)^{1/p}$$

for all x and y in C. Since every continuous function on $[0, 1]$ is bounded, the integral here is finite, and it is easy to check the conditions for this d_p to be a metric. So, for each $p \geq 1$, this d_p is a metric on C. Incidentally, the metric of Example 2.17 can be denoted by d_∞ in analogy with d_∞ in Exercise 2.24.

2.29 *Open balls.* Let $E = \mathbb{R}^2$. Describe the open ball $B(x, r)$, for fixed x and r,
    (a) under d_2 of Exercise 2.24,
    (b) under d_1 of Exercise 2.24,
    (c) under d_∞ of Exercise 2.24,
    (d) under d_2 of Exercise 2.26 with $w_1 = 1$ and $w_2 = 5$.

2.30 *Open balls in C.* For the metric space of Example 2.17, describe $B(x, r)$ for a fixed function x and fixed number $r > 0$. Draw pictures!

2.31 *Product spaces.* Let (E_1, d_1) and (E_2, d_2) be arbitrary metric spaces. Let $E = E_1 \times E_2$ and define, for $x = (x_1, x_2)$ in E and $y = (y_1, y_2)$ in E,

$$d(x, y) = [d_1(x_1, y_1)^2 + d_2(x_2, y_2)^2]^{1/2}.$$

Show that d is a metric on E. The metric space (E, d) is called the product of the metric spaces (E_1, d_1) and (E_2, d_2).

C. Open and Closed Sets

Let (E, d) be a metric space. All points mentioned below are points of E, all sets are subsets of E. Recall the definition 2.21 of the open ball $B(x, r)$ with center x and radius r.

2.32 DEFINITION. A set A is said to be *open* if for every x in A there is a number $r > 0$ such that $B(x, r) \subset A$. A set is said to be *closed* if its complement is open.

For example, if $E = \mathbb{R}$ with the usual distance, the intervals (a, b), $(-\infty, b)$, (a, ∞) are open, the intervals $[a, b]$, $(-\infty, b]$, $[a, \infty)$ are closed, and the interval $(a, b]$ is neither open nor closed.

2.33 PROPOSITION. Every open ball is open.

PROOF. Fix x and r. To show that $B(x, r)$ is open, we need to show that for every y in $B(x, r)$ there is a number $q > 0$ such that $B(y, q) \subset B(x, r)$. This is accomplished by picking $q = r - d(x, y)$. Since y is in $B(x, r)$, we have $d(x, y) < r$ and, hence, $q > 0$. And, every point of $B(y, q)$ is a point of $B(x, r)$, because $z \in B(y, q)$ means $d(z, y) < q$, which implies that

$$d(z, x) \leq d(z, y) + d(y, x) < q + d(y, x) = r.$$

$\square$

2.34 THEOREM. The sets $\emptyset$ and E are open. The intersection of a finite number of open sets is open. The union of an arbitrary collection of open sets is open.

PROOF. The first assertion is trivial from the definition.

We prove the second assertion for the intersection of two open sets. The general case follows from the repeated application of the case for two. Let A and B be open. Let $x \in A \cap B$. Since A is open and x is in A, there is a number $p > 0$ such that $B(x, p) \subset A$. Similarly, there is a number $q > 0$ such that $B(x, q) \subset B$. Let $r = p \wedge q$, the smaller of p and q. Then, $B(x, r) \subset B(x, p) \subset A$ and $B(x, r) \subset B(x, q) \subset B$. Hence, $B(x, r) \subset A \cap B$. So, $A \cap B$ is open.

For the last assertion, let $\{A_i : i \in I\}$ be an arbitrary collection of open sets. We want to show that $A = \bigcup_i A_i$ is open. Let x be in A. Then, $x \in A_i$ for some $i \in I$. Since A_i is open, there is a number $r > 0$ such that $B(x, r) \subset A_i$. Since $A_i \subset A$, this shows that $B(x, r) \subset A$. So, A is open. $\square$

The following characterization is immediate from the preceding theorem together with Proposition 2.33.

2.35 PROPOSITION. A set is open if and only if it is the union of a collection of open balls.

PROOF. If A is the union of a collection of open balls, then A must be open in view of Propositions 2.33 and 2.34. To show the converse, let A be open. Then, for every x in A, there is an open ball $A_x = B(x, r(x))$ contained in A. Obviously, the union of all these A_x is exactly A. $\square$

Closed Sets

Recall that a subset of E is closed if and only if its complement is open. Thus, the following theorem is immediate from Theorem 2.34 above and the fact that the complement of a union is the intersection of complements and vice versa.

2.36 THEOREM. The sets $\emptyset$ and E are closed. The union of finitely many closed sets is closed. The intersection of an arbitrary collection of closed sets is closed.

Every closed ball is closed. This last observation can be proved along the lines of Proposition 2.33: if $y \in E \setminus \bar{B}(x, r)$ then $d(y, x) > r$, and picking $p = d(x, y) - r > 0$, we see that $B(y, p) \subset E \setminus \bar{B}(x, r)$, which proves that $E \setminus \bar{B}(x, r)$ is open. In particular, for each x in E, the singleton $\{x\}$ is closed. It follows from this and the preceding theorem that every finite set is closed.

Interior, Closure, and Boundary

Let A be a subset of E. The collection of all closed sets containing A is not empty (since E belongs to that collection). The intersection $\bar{A}$ of that collection is a closed set by the last theorem. The set $\bar{A}$ is called the *closure* of A. Clearly, $\bar{A}$ is the smallest closed set that contains A, that is, if $B \supset A$ and B is closed then $B \supset \bar{A}$.

We define the *interior* of A similarly as the largest open set contained in A, and we denote it by A°. In other words, A° is the union of all open sets contained in A. Note that

2.37 $$A^\circ \subset A \subset \bar{A}.$$

We define the *boundary* of A to be the set $\partial A = \bar{A} \setminus A^\circ$.

For example, if A is the open ball $B(x, r)$ in the Euclidean space $E = \mathbb{R}^n$, then $A^\circ = A$, $\bar{A} = \bar{B}(x, r)$, and ∂A is the sphere of radius r centered at x. If $E = \mathbb{R}$ with the usual metric, and if $A = (a, b]$, then $\bar{A} = [a, b]$ and $A^\circ = (a, b)$ and $\partial A = \{a, b\}$. The following seems self-evident.

2.38 PROPOSITION. A set is closed if and only if it is equal to its closure. A set is open if and only if it is equal to its interior.

Open Subsets of the Real Line

We take $E = \mathbb{R}$ with the usual distance. Then, every open ball is an open interval, and according to Proposition 2.35, every open set is the union of a collection of open balls. The following sharpens the picture by taking into account the special nature of the real line.

2.39 THEOREM. A subset of $\mathbb{R}$ is open if and only if it is the union of a countable collection of disjoint open intervals.

PROOF. The "if" part is immediate from Proposition 2.35 and the fact that every open ball is an interval in this case.

To prove the "only if" part, let A be an open subset of $\mathbb{R}$. Recall that the set $\mathbb{Q}$ of all rationals is countable. For each q in $\mathbb{Q} \cap A$, let

$$a_q = \sup\{y \leq q : y \notin A\}, \quad b_q = \inf\{y \geq q : y \notin A\}.$$

Then,

$$B = \bigcup_{q \in \mathbb{Q} \cap A} (a_q, b_q)$$

is the union of a countable collection of open intervals. We show next that $A = B$ by showing that $A \subset B$ and $B \subset A$.

Let x be in A. Since A is open, there is a ball $B(x, r)$ contained in A. Take a rational number q in this ball. Clearly, $B(x, r) \subset (a_q, b_q)$. Thus, x is in B. Since this is true for every x in A, we have that $A \subset B$.

Fix $q \in \mathbb{Q} \cap A$. Clearly, $(a_q, b_q) \subset A$. Hence, $B \subset A$.

We have shown that $A = B$, and B has the desired form except that the intervals (a_q, b_q) are not necessarily disjoint. Note that if $r \in (a_q, b_q)$ then $(a_r, b_r) = (a_q, b_q)$ and $q \in (a_r, b_r)$. Let us write $q \approx r$ if and only if $(a_q, b_q) = (a_r, b_r)$. This defines an equivalence relation on the set $\mathbb{Q} \cap A$. Thus, by picking exactly one q from each equivalence class, we can form a set $I \subset \mathbb{Q} \cap A$ such that $(a_q, b_q) \cap (a_r, b_r) = \emptyset$ for all distinct q and r in I, and

$$A = B = \bigcup_{q \in I} (a_q, b_q).$$

$\square$

2.40 EXAMPLE. *The Cantor set.* Start with the unit interval $\mathbb{B} = [0, 1]$. To each q in the set $I = \{1/2;\ 1/4,\ 3/4;\ 1/8,\ 3/8,\ 5/8,\ 7/8;\ 1/16,\ 3/16, \ldots, 15/16; \ldots\}$ we associate an open interval D_q in the following fashion: $D_{1/2}$ is the open interval $(1/3,\ 2/3)$, which is the middle third of $\mathbb{B}$. Deleting it from $\mathbb{B}$ leaves two closed intervals, $[0,\ 1/3]$ and $[1/3,\ 1]$. Let $D_{1/4}$ be the interval $(1/9,\ 2/9)$, which is the middle third of $[0,\ 1/3]$, and let $D_{3/4}$ be $(7/9,\ 8/9)$, which is the middle third of $[2/3,\ 1]$. Deleting those middle thirds, we are left with four closed intervals of length $1/9$ each. Let $D_{1/8}, D_{3/8}, D_{5/8}, D_{7/8}$ be the open intervals that make up the middle

$$0 \qquad\qquad\qquad\qquad\qquad\qquad\qquad\qquad\qquad\qquad\qquad 1$$

FIGURE 2.1. The set $D = \bigcup D_q$.

thirds of those closed intervals. Delete the middle thirds, and continue in this manner (see Fig. 2.1). Then,

$$D = \bigcup_{q \in I} D_q$$

is the union of the countably many disjoint open intervals D_q, $q \in I$. It is an example of a nontrivial open set. Incidentally, note that the lengths of the D_q sum to

$$\frac{1}{3} + \left(\frac{1}{9} + \frac{1}{9}\right) + \left(\frac{1}{27} + \frac{1}{27} + \frac{1}{27} + \frac{1}{27}\right) + \cdots = 1.$$

Thus, the "length" of D is 1. But the set $C = \mathbb{B} \setminus D$ is not empty.

The set $C = \mathbb{B} \setminus D$ is called the *Cantor set*. It is a closed set. The construction above shows that C is obtained by starting with $\mathbb{B}$ and deleting the middle third of every interval we can find. Thus, there is no open interval contained in C. That is, there are no open balls in C. Hence, the interior of C must be empty, and C is pure boundary:

$$C^\circ = \emptyset, \quad \bar{C} = C, \quad \partial C = C.$$

Also, since the length of D is equal to the length of $\mathbb{B}$, the length of $C = \mathbb{B} \setminus D$ must be 0. In summary, the Cantor set is very thin.

Nevertheless, C has at least as many points as the interval $[0, 1]$. We prove this next by showing, via construction, that there exists an injection g from $[0, 1]$ into C.

To this end, we start by defining an increasing function f from D into $[0, 1]$ by letting

$$f(x) = q, \quad \text{if } x \in D_q.$$

Then, we define the function g on $[0, 1]$ by setting $g(1) = 1$ and

$$g(y) = \inf\{x \in D : f(x) > y\}, \quad 0 \le y < 1.$$

We show first that $g(y) \in C$ for every y. This is obvious for $y = 1$. Let $y \in [0, 1)$; note that $g(y)$ is the infimum of the union of all intervals D_q with $q > y$; that infimum cannot belong to D; so $g(y)$ must belong to C (since it is obvious that $g(y) \in \mathbb{B}$). Finally, we show that $g \colon [0, 1] \mapsto C$ is an injection by showing that if $y < z$, then $g(y) < g(z)$. Fix $y < z$. Note that there is at least one q in I such that $y < q < z$, and the corresponding set D_q is contained in $\{x \in D : f(x) > y\}$ but not in $\{x \in D : f(x) > z\}$. It follows that the number $g(y)$ is to the left of the interval D_q whereas $g(z)$ is to the right. So, $g(y) < g(z)$ if $y < z$. Hence, $g \colon [0, 1] \mapsto C$ is an injection (Fig. 2.2).

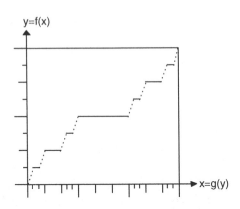

FIGURE 2.2. The Cantor function f.

Exercises and Complements

2.41 Let (E, d) be a metric space. Show that

$$
\begin{aligned}
\bar{A} &= \{x \in E : d(x, A) = 0\}, \\
A^\circ &= \{x \in E : d(x, A^c) > 0\}, \\
\partial A &= \{x \in E : d(x, A) = 0 \text{ and } d(x, A^c) = 0\}.
\end{aligned}
$$

2.42 Let (E, d) be a metric space. Fix $A \subset E$. Let $A_\epsilon = \{x \in E : d(x, A) < \epsilon\}$ for each $\epsilon > 0$. Show that A_ϵ is an open set containing A for each $\epsilon > 0$. Show that $\bar{A} = \bigcap_{\epsilon > 0} A_\epsilon$.

2.43 *Boundedness.* Let (E, d) be a metric space. Show that a subset A of E is bounded if and only if it is contained in some ball, that is, if and only if $A \subset B(x, r)$ for some x and r.

2.44 Take $E = \mathbb{R}$ and d the usual metric. Let $A \subset E$. Show that if A is closed and bounded above, then $\sup A$ belongs to A (that is, A has a maximum). Similarly, if A is closed and bounded below, then it has a minimum. Show that an open set A cannot have a minimum, that is, $\inf A$ cannot belong to A.

2.45 Let D be the open set of Example 2.40. Find its interior and boundary.

2.46 *Denseness.* A set D is said to be *dense* in E if $\bar{D} = E$. Let D be dense in E. Show that every x in E is at 0 distance from D. Thus, every open ball has at least one point of D. Show that the set $\mathbb{Q}$ of all rationals is dense in $\mathbb{R}$, the set of all pairs of rationals is dense in $\mathbb{R}^2$, etc.

2.47 *Separability.* A metric space E is said to be separable if there exists a countable set D that is dense in E. So, for example, the Euclidean spaces $\mathbb{R}$, $\mathbb{R}^2$, $\mathbb{R}^3$, ... are separable.

2.48 *Discrete metric spaces.* Let E be arbitrary and suppose that d is the discrete metric (see Exercise 2.23 for this) on E. Show that each subset A is both open and closed. For $r \leq 1$, every open ball $B(x, r)$ consists of exactly the point x. Note that $B(x, 1) = \{x\}$, $\bar{B}(x, 1) = E$ for every x. (Moral: $\bar{B}(x, r)$ is not necessarily the closure of $B(x, r)$.) If E is countable, then it is separable (trivially). If E is uncountable, it is not separable. Show this.

2.49 *Half-spaces.* A set H in $\mathbb{R}^n$ of the form $H = \{x : \xi \cdot x \leq b\}$, where ξ in $\mathbb{R}^n$ and b in $\mathbb{R}$ are given, is called a half-space. Show that half-spaces are closed.

D. Convergence

Let (E, d) be a metric space. Our goal is to discuss the notion of convergence for a sequence of points in E. We do so by employing the concept of convergence in $\mathbb{R}$, for which we refer to Sect. D of Chap. 1.

2.50 DEFINITION. A sequence (x_n) in E is said to be *convergent* in E if there exists a point x in E such that $\lim d(x_n, x) = 0$. If so, then (x_n) is said to *converge* to x, the point x is called the *limit* of (x_n), and the notation $x = \lim x_n$ is used to indicate this.

2.51 REMARK. The preceding definition includes, implicitly within it, the fact that a convergent sequence has exactly one limit. To see this, suppose that (x_n) converges to x and to y, that is, $\lim d(x_n, x) = 0$ and $\lim d(x_n, y) = 0$. Then,

$$0 \leq d(x, y) \leq d(x, x_n) + d(x_n, y)$$

by the triangle inequality, and the right side converges to zero. Thus, $d(x, y) = 0$, which means that $x = y$.

The following brings together a number of rewordings of convergence. Each is a slight alteration of the others. No proof seems needed.

2.52 THEOREM. The following statements are equivalent:

(a) (x_n) converges to x.
(b) For every $\epsilon > 0$ there is an integer n_ϵ such that $d(x_n, x) < \epsilon$ for all $n \geq n_\epsilon$.
(c) The set $\{n : d(x_n, x) \geq \epsilon\}$ is finite for each $\epsilon > 0$.
(d) For every $\epsilon > 0$, the ball $B(x, \epsilon)$ includes all but a finite number of the terms x_n.

2.53 COROLLARY. Every convergent sequence is bounded.

PROOF. Let (x_n) be convergent and x its limit. In view of the equivalence of (a) and (d) in the preceding theorem, $B(x, 1)$ includes all but a finite number of the terms x_n. Let r be the maximum of the distances from x to those terms x_n outside $B(x, 1)$, if there are any; otherwise, set $r = 1$. Clearly $r < \infty$ and $B(x, r)$ contains (x_n), which means that (x_n) is bounded.   □

Subsequences

It follows from Theorem 2.52 that we may remove a finite number of terms, or rearrange the terms, without affecting the convergence. The following generalizes this.

2.54 PROPOSITION. If a sequence converges to x, then every subsequence of it converges to the same x.

PROOF. Let (x_n) be a sequence with limit x. Let (y_n) be a subsequence of it, that is, $y_n = x_{k_n}$ for some $k_1 < k_2 < \cdots$. Now, by Theorem 2.52, for every $\epsilon > 0$, the ball $B(x, \epsilon)$ includes all the terms x_n except for some finite number of them; therefore the same must be true for the terms y_n. So, by Theorem 2.52, the subsequence (y_n) converges to x.   □

Convergence and Closed Sets

Think of a particle that moves in E by jumps: first it is at x_1, then at x_2, then at x_3, and so on. The following gives meaning to the term "closed set" if you think of sequences in this fashion.

2.55 THEOREM. A set is closed if and only if it includes the limit of every convergent sequence in it.

PROOF. *"Only if" part.* Suppose that A is a closed set and that (x_n) is a sequence in A with limit x. We show that, then, x must belong to A. For, otherwise, if x were in A^c, there would exist an $\epsilon > 0$ such that $B(x, \epsilon) \subset A^c$ since A^c is open, and $B(x, \epsilon)$ would include infinitely many terms since x is the limit, which would contradict the fact that all the x_n are in A.

   "If" part. We show that if A is not closed then there is a sequence (x_n) in A that converges to some point x in A^c. Suppose that A is not closed. Then A^c is not open. Thus, there exists an x in A^c such that $B(x, r) \cap A$ has at least one point for each $r > 0$. Hence, for each n in $\mathbb{N}^*$, there is an x_n in A such that $d(x_n, x) < 1/n$. Obviously, (x_n) is in A and converges to x which is not in A.   □

Exercises

2.56 *Discrete metric spaces.* Suppose that d is the discrete metric on E. Show that (x_n) is convergent if and only if it is ultimately *stationary*, that is, if and only if it has the form $(x_1, x_2, \ldots, x_n, x, x, x, \ldots)$ for some n.

2.57 Let (E, d) be a metric space. Show that if (x_n) converges to x and (y_n) converges to y, then $d(x_n, y_n)$ converges to $d(x, y)$. Hint: first show that, for arbitrary x, y, z in E,

$$|d(x, y) - d(x, z)| \leq d(y, z).$$

Use this to write

$$\begin{aligned}|d(x_n, y_n) - d(x, y)| &\leq |d(x_n, y_n) - d(x_n, y)| + |d(x_n, y) - d(x, y)| \\ &\leq d(y_n, y) + d(x_n, x),\end{aligned}$$

and take limits.

2.58 Show that if (x_n) converges to x, then $d(x_n, A)$ converges to $d(x, A)$ for each fixed subset A of E.

E. Completeness

   Let (E, d) be a metric space. Recall that a sequence (x_n) in E is convergent if there is an x in E such that $\lim d(x_n, x) = 0$. This definition has two shortcomings. First, starting with (x_n), we rarely have a candidate x for the limit. Second, often we are not interested in computing the limit itself; it is generally sufficient to know that the limit exists and has such and such properties. This section is aimed at rectifying these shortcomings.

Cauchy Sequences

2.59 DEFINITION. A sequence (x_n) in E is said to be *Cauchy* if for every $\epsilon > 0$ there is an integer n_ϵ such that $d(x_m, x_n) < \epsilon$ for all $m > n \geq n_\epsilon$.

The following is nearly a restatement of this definition in more geometric terms.

2.60 LEMMA. A sequence (x_n) is Cauchy if and only if for every $\epsilon > 0$ there is a ball of radius ϵ that contains all but finitely many of the terms x_n.

PROOF. Suppose that (x_n) is Cauchy. Let $\epsilon > 0$. Then, there is n_ϵ such that $d(x_m, x_n) < \epsilon$ for all $m > n \geq n_\epsilon$. Thus, in particular, the ball $B(x_{n_\epsilon}, \epsilon)$ contains all the terms except possibly $x_1, \ldots, x_{n_\epsilon - 1}$. This proves the necessity of the condition.

Conversely, suppose that for every $\epsilon > 0$ there is a ball $B(x, \epsilon)$ with some x as its center such that all but a finite number of the terms are in the ball. Given $\epsilon > 0$, pick x so that $B(x, \epsilon/2)$ contains all the x_n except perhaps finitely many, that is, there is n_ϵ such that $x_n \in B(x, \epsilon/2)$ for all $n \geq n_\epsilon$. Now, if $m > n \geq n_\epsilon$, then

$$d(x_m, x_n) \leq d(x_m, x) + d(x, x_n) < \epsilon/2 + \epsilon/2 = \epsilon.$$

Hence, (x_n) is Cauchy. This proves the sufficiency.   □

2.61 THEOREM.

(a) Every convergent sequence is Cauchy.
(b) Every Cauchy sequence is bounded.
(c) Every subsequence of a Cauchy sequence is Cauchy.

PROOF. The first claim is immediate from the preceding lemma and Theorem 2.52. The second claim is proved, via the preceding lemma, by following the proof of Corollary 2.53. The last claim is immediate from the preceding lemma.   □

The following shows that if a sequence is Cauchy and one can find a subsequence of it that converges to some point x, then the original sequence converges to x.

2.62 PROPOSITION. A Cauchy sequence that has a convergent subsequence is itself convergent.

PROOF. Let (x_n) be Cauchy. Let x be the limit of a convergent subsequence of it. Pick $\epsilon > 0$. By Lemma 2.60, there is a ball $B(y, \epsilon)$ that contains all but a finite number of the x_n. That ball $B(y, \epsilon)$ must contain all but a finite number of the subsequence as well. Thus, x must be in $\bar{B}(y, \epsilon)$. Then, $B(x, 3\epsilon)$ contains $\bar{B}(y, \epsilon)$ and hence contains all but a finite number of the x_n. Thus, (x_n) is convergent and $x = \lim x_n$ in view of Theorem 2.52.   □

Complete Metric Spaces

The results above suggest that all Cauchy sequences should be convergent. Unfortunately, this is not true in general. Here is an example.

Suppose that $E = \mathbb{Q}$, the set of all rationals, with the metric it inherits from the real line. Let $x = \sqrt{2}$, which is not a rational number, and let (x_n) be a sequence in $\mathbb{Q}$ that converges to x in the sense of convergence in $\mathbb{R}$: for instance, pick x_n to be a rational number in the interval $(x, x + 1/n)$ for each n. Over the metric space $\mathbb{Q}$, the sequence (x_n) is Cauchy, but fails to be convergent in $\mathbb{Q}$ simply because x is not in $\mathbb{Q}$. The problem here is not with the Cauchy sequence, but with the space $\mathbb{Q}$. The space $\mathbb{Q}$ has holes in it!

The following introduces the extra notion we want.

2.63 DEFINITION. The metric space E is said to be *complete* if every Cauchy sequence in E converges to a point of E.

Theorem 2.61a becomes stronger: when E is complete, a sequence is convergent if and only if it is Cauchy. The following is immediate from Theorem 2.55.

2.64 PROPOSITION. If E is complete and D is a closed subset of E, then D is a complete metric space with the metric it inherits from E.

The next theorem shows that certain familiar spaces are complete. Other examples are listed in exercises.

2.65 THEOREM. Every Euclidean space is complete.

PROOF. We start with the one-dimensional Euclidean space, namely $\mathbb{R}$. Let $(x_n) \subset \mathbb{R}$ be Cauchy. Then, for every $\epsilon > 0$ there is a ball of radius ϵ (namely an open interval of length 2ϵ) that contains all but finitely many of the x_n. Therefore, the numbers $x = \liminf x_n$ and $y = \limsup x_n$ must belong to that ball, which means that $0 \leq y - x < 2\epsilon$. Since this is true for every $\epsilon > 0$, we must have $x = y$, that is, (x_n) is convergent. This proves that $\mathbb{R}$ is complete.

Now, fix $k \geq 2$ and consider the Euclidean space $\mathbb{R}^k$. We write $x = (a, b, \ldots, c)$ for each x in $\mathbb{R}^k$ for simplicity of notation (in other words, the coordinates of x are $a, b, \ldots, c$).

Consider a Cauchy sequence of points $x_n = (a_n, b_n, \ldots, c_n)$ in $\mathbb{R}^k$. Given $\epsilon > 0$, then, for all m and n large enough, we have

$$d(x_m, x_n) = (|a_m - a_n|^2 + |b_m - b_n|^2 + \cdots + |c_m - c_n|^2)^{1/2} < \epsilon,$$

which shows that

$$|a_m - a_n| < \epsilon, \quad |b_m - b_n| < \epsilon, \quad \ldots, \quad |c_m - c_n| < \epsilon.$$

In other words, the sequences $(a_n), (b_n), \ldots, (c_n)$ in $\mathbb{R}$ are Cauchy. We have just shown that $\mathbb{R}$ is complete. So, these sequences must be convergent in $\mathbb{R}$, say, with limits $a, b, \ldots, c$ respectively. Now, let $x = (a, b, \ldots, c)$ and note that

$$d(x_n, x)^2 = |a_n - a|^2 + |b_n - b|^2 + \cdots + |c_n - c|^2$$

converges to 0. Hence, $\lim d(x_n, x) = 0$, and (x_n) is convergent. This completes the proof that $\mathbb{R}^k$ is complete.   □

Exercises and Complements

2.66   Show that the following metric spaces are complete:

(a) $E = \mathbb{R}^2$ with the Manhattan metric d.

(b) E arbitrary, d is the discrete metric.

In fact, each $\mathbb{R}^n$ is a complete metric space with every one of the metrics introduced in Exercises 2.24 and 2.26.

2.67   Show that the space l^2 introduced in Exercise 2.27 is complete. Incidentally, so is the space $\mathcal{C}$ of Example 2.17 and Exercise 2.28.

2.68 Two Cauchy sequences (x_n) and (y_n) are said to be equivalent if their merger $(x_1, y_1, x_2, y_2, \ldots)$ is Cauchy. In this case, we write $(x_n) \equiv (y_n)$. Show that this defines an equivalence relation. That is,

    (a) $(x_n) \equiv (x_n)$,
    (b) $(x_n) \equiv (y_n)$ implies that $(y_n) \equiv (x_n)$,
    (c) $(x_n) \equiv (y_n)$, $(y_n) \equiv (z_n)$ implies that $(x_n) \equiv (z_n)$.

F. Compactness

Let (E, d) be a metric space. It will be convenient to refer to E as a metric space, without mentioning d. Also, all sets mentioned will be subsets of E. We shall use the picturesque phrase "the collection $\{A_i : i \in I\}$ covers B" to mean that $\bigcup_{i \in I} A_i \supset B$.

2.69 DEFINITION. A set C is said to be *compact* if every collection of open sets that covers C has a finite subcollection that covers C. The metric space (E, d) is said to be compact if E is so.

We shall show that, for many metric spaces, compact sets are precisely the sets that are bounded and closed. The following are aimed in that direction.

2.70 PROPOSITION. Every compact set is bounded.

PROOF. Let C be compact. Cover C by the collection of balls of radius 1 about each x in C. By the definition of compactness, C can be covered by finitely many of these balls. Since a finite union of bounded sets is bounded, C is bounded.   □

2.71 PROPOSITION. Every closed subset of a compact set is compact.

PROOF. Let D be compact. Let $C \subset D$ be closed. Fix a collection of open sets that covers C. Adding the open set $E \setminus C$ to that collection, we obtain a collection of open sets that covers D. Since D is compact, the latter collection has a finite subcollection that still covers D. Removing $E \setminus C$ from that subcollection (if it were in), we obtain a finite subcollection of the original collection that covers C. Thus, C must be compact.   □

Compact Subspaces

Recall that every subset D of E can be regarded as a metric space by itself, with the metric it inherits from E; see Proposition 2.64. Whether D is open or not as a subset of E, it is open automatically when it is regarded as a metric space. Compactness is not so shiftless.

2.72 PROPOSITION. A set D is compact as a metric space if and only if it is compact as a subset of E.

PROOF. A subset of D is an open ball in the space D if and only if it has the form $B \cap D$ for some open ball B of the space E. Since an open set is the union of all the open balls it contains, it follows that A is an open subset of the space D if and only if $A = B \cap D$ for some open subset B of the space E. Now, the definition of compactness does the rest. □

Cluster Points, Convergence, Completeness

This is to look into the connections between compactness and convergence.

2.73 DEFINITION. A point x in E is called a *cluster point* of a subset A of E provided that every open ball centered at x contains infinitely many points of A.

2.74 THEOREM. Every infinite subset of a compact set has at least one cluster point in that compact set.

PROOF. We shall show that if C is compact, and $A \subset C$, and A has no cluster point in C, then A is finite. Let A and C be such. Since no x in C is a cluster point of A, for every x in C there is an open ball $B(x, r)$ that contains only finitely many points of A. Those open balls cover C obviously. Since C is compact, there must be a finite number of them that cover C and, therefore, A. Since each one of those finitely many balls has a finite number of points of A, the total number of points in A must be finite. □

The following theorem shows how compactness helps in discussing convergence. In particular, together with Proposition 2.62, it shows that every Cauchy sequence in a compact set is convergent.

2.75 THEOREM. Every sequence in a compact set has a subsequence that converges to some point of that set.

PROOF. Let C be compact. Let $(x_n) \subset C$. If the set $A = \{x_1, x_2, \dots\}$ is finite, then at least one point of A, say x, appears infinitely often in the sequence, and hence $(x, x, \dots)$ is a subsequence which obviously converges to $x \in A \subset C$. Now suppose that A is infinite. By the preceding theorem, then A has a cluster point x in C. Since each one of the balls $B(x, 1/n)$, $n \in \mathbb{N}^*$, has infinitely many points in C, we may pick k_1 so that x_{k_1} is in $B(x, 1)$, pick $k_2 > k_1$ so that x_{k_2} is in $B(x, 1/2)$, pick $k_3 > k_2$ so that x_{k_3} is in $B(x, 1/3)$, and so on. Obviously, (x_{k_n}) converges to x. □

2.76 COROLLARY. Every compact set is closed.

PROOF. Let C be compact. The preceding theorem implies that every convergent sequence in C converges to a point of C. Thus, C is closed by Theorem 2.55. □

2.77 COROLLARY. Every compact metric space is complete. Every Cauchy sequence in a compact metric space is convergent.

PROOF. The second statement is immediate from Theorem 2.75 and Proposition 2.62. The first follows from the second by the definition of completeness. □

Compactness in Euclidean Spaces

We have seen that, for an arbitrary metric space, every compact set is bounded and closed; see Proposition 2.70 and Corollary 2.76. In the case of Euclidean spaces, the converse is true as well. This is called the *Heine–Borel theorem.*

2.78 THEOREM. A subset of a Euclidean space is compact if and only if it is bounded and closed.

We start by listing an auxiliary result that is trivial at least for $\mathbb{R}$, $\mathbb{R}^2$, $\mathbb{R}^3$. We omit its proof.

2.79 LEMMA. Let B be a bounded subset of a Euclidean space E. Then, for every $\epsilon > 0$ there is a finite collection of closed balls of radius ϵ that covers B.

PROOF OF THEOREM 2.78. As mentioned above, Proposition 2.70 and Corollary 2.76 prove the necessity part. We now prove the sufficiency of the condition.

Let E be a Euclidean space and let C be a closed and bounded subset of E. Suppose that C is not compact. Then, there is a collection $\{A_i : i \in I\}$ of open sets that covers C but is such that

2.80   no finite subcollection of $\{A_i : i \in I\}$ covers C.

(a) Let $\epsilon = 1/2$. By the preceding lemma, we can find a finite number m of closed balls $B_1, \ldots, B_m$ of radius ϵ that cover C. Then, $C = (C \cap B_1) \cup \cdots \cup (C \cap B_m)$. In view of 2.80, at least one of $C \cap B_1$, ..., $C \cap B_m$ cannot ever be covered by a finite subcollection of the A_i; let that one be denoted by C_1. Now, C_1 is closed, its diameter is at most $2\epsilon = 1$ (since the B_k have diameter 1), and 2.80 is true for C_1.

(b) Applying the arguments of the preceding paragraph with $\epsilon = 1/4$ to the set C_1 we get a new set $C_2 \subset C_1$ that is closed, has diameter at most $1/2$, and 2.80 holds for C_2. Repeating this with $\epsilon = 1/6, 1/8, 1/10, \ldots$ we obtain further sets $C_3, C_4, C_5, \ldots$ with the same properties but with diameters at most $1/3, 1/4, 1/5$, ... Clearly $C_1 \supset C_2 \supset C_3 \supset \cdots$.

(c) Since 2.80 holds for each C_n, it must be that no C_n is empty (covering an empty set takes no effort). Thus, we may pick x_1 from C_1 and x_2 from C_2, and so on to obtain a sequence (x_n).

(d) This sequence is Cauchy: given $\epsilon > 0$ choose n so that $1/2n < \epsilon$, and then $x_n, x_{n+1}, \ldots$ are all in a ball of radius ϵ since all these terms are in C_n which has diameter less than $1/n$. Since E is Euclidean, it is complete (see Theorem 2.65), which means that every Cauchy sequence converges. Hence, the sequence (x_n) converges to some point x_0 in E. Since, for each n, $(x_m : m \geq n) \subset C_n$ and C_n is closed, the limit x_0 belongs to C_n by Theorem 2.55.

(e) Since the A_i cover C, there must exist an index i in I such that x_0 is in A_i. Fix that i. Since A_i is open, there is a number $\epsilon > 0$ such that

$$B(x_0, \epsilon) \subset A_i.$$

Now choose n large enough that $1/n < \epsilon/2$. Since $x_0 \in C_n$ and $\operatorname{diam} C_n \leq 1/n < \epsilon/2$, we see that

$$C_n \subset B(x_0, \epsilon).$$

In other words, A_i covers C_n. This contradicts the earlier assertion that 2.80 holds for all C_n. This completes the proof.   □

Exercises

2.81 *Supremums.* Let A be a nonempty subset of $\mathbb{R}$. Suppose that A is bounded above but has no greatest element. Show that, then, $\sup A$ is a cluster point of A.

2.82 Show that the union of a finite number of compact sets is again compact.

2.83 Give an example of an infinite subset of $\mathbb{R}$ that has no cluster points. Give an example of one with exactly two cluster points. Identify the cluster points of the set

$$ A = \left\{ x \in \mathbb{R} : x = \frac{1}{m} + \frac{1}{n} \text{ for some } m, n \text{ in } \mathbb{N}^* \right\}. $$

2.84 *Sequences in* $\mathbb{R}$. By 2.78, the Heine–Borel theorem, every closed interval $[a, b] \subset \mathbb{R}$ is compact. Thus, every bounded sequence in $\mathbb{R}$ has a convergent subsequence (cf. Theorem 2.75). Another consequence is the following useful result:

Let (x_n) be a bounded sequence in $\mathbb{R}$. Suppose that all convergent subsequences of it have the same limit x. Then, (x_n) converges to x.

Prove this by following the steps below.

(a) Show that $\underline{x} = \liminf x_n$ and $\bar{x} = \limsup x_n$ are cluster points of (x_n).

(b) Show that there is a subsequence of (x_n) that converges to $\underline{x}$. Similarly, then, there is a subsequence that converges to $\bar{x}$.

(c) By the hypothesis that all convergent subsequences have the same limit, we conclude that $\underline{x} = \bar{x}$, which means that $\lim x_n$ exists (and is in $\mathbb{R}$ since (x_n) is bounded).

CHAPTER 3

Functions on Metric Spaces

Elementary analysis is mostly about functions from $\mathbb{R}$ into $\mathbb{R}$, functions from $\mathbb{R}^n$ into $\mathbb{R}$, or, somewhat more generally, functions from $\mathbb{R}^n$ into $\mathbb{R}^m$. Our aim is to consider functions from one metric space to another. Replacing Euclidean spaces by metric spaces introduces no new difficulties and is useful for dealing with various problems concerning differential and integral equations.

For mappings from one metric space to another we employ either notations like T, S, U or notations like f, g, h. Generally, the transformation notation is cleaner: we write Tx for the image of x under T, which becomes $f(x)$ in the standard function notation.

A. Continuous Mappings

Throughout this section, E, E', etc. will be metric spaces with corresponding metrics d, d', etc. Given a mapping T from E into E', we write Tx for the image of the point x of E and $T^{-1}B$ for the inverse image of the subset B of E'. On a first reading, the reader may wish to take $E' = \mathbb{R}$ and $d'(x,y) = |x-y|$ as usual.

3.1 DEFINITION. A mapping $T \colon E \mapsto E'$ is said to be *continuous at the point x of E* provided that for every $\epsilon > 0$ there is a number $\delta > 0$ such that

$$y \in E, \quad d(x,y) < \delta \quad \Rightarrow \quad d'(Tx, Ty) < \epsilon.$$

The mapping T is said to be *continuous* if it is continuous at every x of E.

3.2 REMARKS. (a) In the definition, δ is allowed to depend on ϵ and x.

(b) When $E = E' = \mathbb{R}$ with the usual metric, the preceding is the classic definition of continuity.

(c) The condition for T to be continuous at x can be rephrased in more geometric terms as follows: for every $\epsilon > 0$ there is a number $\delta > 0$ such that T maps the open ball $B(x, \delta)$ of E into the open ball $B'(Tx, \epsilon)$ of E'. Here,

$$B(x,\delta) = \{y \in E : d(x,y) < \delta\}, \quad B'(Tx,\epsilon) = \{y \in E' : d'(Tx,y) < \epsilon\}.$$

E. Çınlar and R.J. Vanderbei, *Real and Convex Analysis*, Undergraduate Texts in Mathematics, 47
DOI 10.1007/978-1-4614-5257-7_3, © Springer Science+Business Media New York 2013

Continuity and Open Sets

3.3 THEOREM. A mapping $T\colon E \mapsto E'$ is continuous if and only if $T^{-1}B$ is an open subset of E for every open subset B of E'.

PROOF. Suppose that T is continuous. Let $B \subset E'$ be open. We want to show that $A = T^{-1}B$ is open, that is, for every x in A there is $\delta > 0$ such that $B(x,\delta) \subset A$. To this end, fix x in A, note that $y = Tx$ is in B, and, therefore, there is $\epsilon > 0$ such that $B'(y,\epsilon) \subset B$ (because B is open). By the continuity of T, for that ϵ, there is $\delta > 0$ such that T maps $B(x,\delta)$ into $B'(y,\epsilon)$. Because $B'(y,\epsilon) \subset B$, we have $B(x,\delta) \subset A$ as needed.

Suppose that $T^{-1}B$ is open in E for every open subset B of E'. Let x in E be arbitrary. We want to show that T is continuous at x. To this end, fix $\epsilon > 0$. Because $B'(Tx,\epsilon)$ is open, its inverse image is open, that is, $A = T^{-1}B'(Tx,\epsilon)$ is an open subset of E. Note that x is in A; therefore, there is a $\delta > 0$ such that $B(x,\delta) \subset A$, and then T maps $B(x,\delta)$ into $B'(Tx,\epsilon)$. So, T is continuous at x. $\qquad\square$

Continuity and Convergence

If (x_n) is a sequence in E, we write $x_n \xrightarrow{d} x$ to mean that (x_n) converges to x in E in the metric d, that is, $d(x_n, x) \to 0$. Similarly, $y_n \xrightarrow{d'} y$ means that the sequence (y_n) in E' converges to y in the metric d'. The following is probably the most useful characterization of continuity.

3.4 THEOREM. A mapping $T\colon E \mapsto E'$ is continuous at the point x of E if and only if

$$(x_n) \subset E, \quad x_n \xrightarrow{d} x \quad \Rightarrow \quad Tx_n \xrightarrow{d'} Tx.$$

PROOF. Suppose that T is continuous at x. Let $(x_n) \subset E$ and suppose that $x_n \xrightarrow{d} x$. We want to show that, then, $Tx_n \xrightarrow{d'} Tx$, which is equivalent to showing that for every $\epsilon > 0$ the ball $B'(Tx,\epsilon)$ contains all but finitely many of the points Tx_n. To this end, fix $\epsilon > 0$. By the continuity of T at x, there is $\delta > 0$ such that T maps $B(x,\delta)$ into $B'(Tx,\epsilon)$. Since $x_n \in B(x,\delta)$ for all but finitely many n, it follows that $Tx_n \in B'(Tx,\epsilon)$ for all but finitely many n, which is as desired.

Suppose that T is not continuous at x. Then, there is $\epsilon > 0$ such that for every $\delta > 0$ there is y in E such that $d(x, y) < \delta$ and $d'(Tx, Ty) \geq \epsilon$. Thus, for that ϵ, taking $\delta = 1, 1/2, 1/3, \ldots$ we can pick $y = x_1, x_2, x_3, \ldots$ such that $d(x_n, x) < 1/n$ and $d'(Tx_n, Tx) \geq \epsilon$. Hence, there is a sequence (x_n) in E such that $x_n \xrightarrow{d} x$ but (Tx_n) does not converge to Tx.   □

Compositions

The following result is recalled best by the phrase "a continuous function of a continuous function is continuous." Here, E, E', and E'' are metric spaces with their own metrics.

3.5 THEOREM. If $T\colon E \mapsto E'$ is continuous at $x \in E$ and $S\colon E' \mapsto E''$ is continuous at $Tx \in E'$, then $S \circ T\colon E \mapsto E''$ is continuous at $x \in E$. If T is continuous and S is continuous, then $S \circ T$ is continuous.

PROOF. The second assertion is immediate from the first. To show the first, let $(x_n) \subset E$ be such that $x_n \xrightarrow{d} x$. If T is continuous at x, then $Tx_n \xrightarrow{d'} Tx$ by the last theorem; and if S is continuous at Tx, this in turn implies that $S(Tx_n) \xrightarrow{d''} S(Tx)$ by the last theorem again, which means that $S \circ T$ is continuous at x.   □

Examples

3.6 *Constants.* Let $T\colon E \mapsto E'$ be defined by $Tx = b$ where b in E' is fixed. This T is continuous.

3.7 *Identity.* Let $T\colon E \mapsto E$ be defined by $Tx = x$. This T is continuous, as is easy to see from Theorems 3.3 or 3.4.

3.8 *Restrictions.* Let $T\colon E \mapsto E'$ be continuous. For D contained in E, the restriction of T to D is the mapping $S\colon D \mapsto E'$ defined by putting $Sx = Tx$ for each $x \in D$. Obviously, the continuity of T implies that of S.

3.9 *Discontinuity.* Let $f\colon \mathbb{R} \mapsto \mathbb{R}$ be defined by setting $f(x) = 1$ if x is rational and $f(x) = 0$ if x is irrational. This function is discontinuous at every x in $\mathbb{R}$. To see this, fix x in $\mathbb{R}$. For every $\delta > 0$, the ball $B(x, \delta)$ has infinitely many rationals and infinitely many irrationals. Thus, it is impossible to satisfy the condition for continuity at x (for $\epsilon < 1$).

3.10 *Lipschitz continuity.* A mapping $T\colon E \mapsto E'$ is said to satisfy a Lipschitz condition if there exists a constant K in $(0, \infty)$ such that

$$d'(Tx, Ty) \leq K d(x, y)$$

for all x, y in E. Every such mapping is continuous: given $\epsilon > 0$, choose $\delta = \epsilon/K$ no matter what x is.

3.11 *Coordinate mappings.* Let $E = \mathbb{R}^n$, the n-dimensional Euclidean space, fix i in $\{1, \ldots, n\}$, and define $P_i\colon \mathbb{R}^n \mapsto \mathbb{R}$ by $P_i x = x_i$, the ith coordinate of x. Then, P_i satisfies the Lipschitz condition above with $K = 1$ and, thus, is continuous.

Real-Valued Functions

Functions f from a metric space E into $\mathbb{R}$ can be combined through arithmetic operations to obtain new functions. For instance, $f + g$ is the function whose value at x is $f(x) + g(x)$. In defining f/g, however, one must exercise some caution at points x where $g(x) = 0$. It is best to limit the definition of f/g to the set $\{x \in E : g(x) \neq 0\}$. The following is immediate from Theorem 3.4.

3.12 PROPOSITION. If $f\colon E \mapsto \mathbb{R}$ and $g\colon E \mapsto \mathbb{R}$ are continuous, then so are $f + g$, $f - g$, $f \cdot g$, and f/g, except that, in the last case, f/g should be treated as a function on $\{x : g(x) \neq 0\}$.

$\mathbb{R}^n$-Valued Functions

These are functions from a metric space E into the Euclidean space $\mathbb{R}^n$ (with the Euclidean distance). The following reduces the notion of continuity for such mappings to the case of real-valued functions. We use the projection mappings P_i introduced in Example 3.11: $P_i x$ is the i-coordinate of the vector x in $\mathbb{R}^n$.

3.13 PROPOSITION. A mapping $T\colon E \mapsto \mathbb{R}^n$ is continuous if and only if the mappings $P_1 \circ T, \ldots, P_n \circ T$ from E into $\mathbb{R}$ are continuous.

PROOF. Let T be continuous. Then, $P_i \circ T$ is continuous for each i because a continuous function of a continuous function is continuous.

Suppose that $P_1 \circ T, \dots, P_n \circ T$ are continuous. To show that, then, T is continuous, we start by observing that

3.14
$$\|u - v\| = \left(\sum_1^n |P_i u - P_i v|^2 \right)^{1/2}, \quad u, v \in \mathbb{R}^n.$$

Now, fix x in E and $\epsilon > 0$. Using the definition of continuity for $P_i \circ T$ at x with $\epsilon_i = \epsilon/\sqrt{n}$, we find $\delta_i > 0$ such that

$$d(x, y) < \delta_i \Rightarrow |P_i T x - P_i T y| < \epsilon/\sqrt{n}.$$

Let $\delta = \min\{\delta_1, \dots, \delta_n\}$. Then $\delta > 0$ and

$$d(x, y) < \delta \quad \Rightarrow \quad |P_i T x - P_i T y| < \epsilon/\sqrt{n} \text{ for each } i$$
$$\Rightarrow \quad \|Tx - Ty\| < \epsilon$$

in view of 3.14 used with $u = Tx$ and $v = Ty$. $\qquad\qquad\qquad\qquad\qquad\qquad \square$

Exercises

3.15 *Continuity of metrics.* For the metric d on E, show that the mapping $x \mapsto d(x, y)$ is continuous for each fixed y. By symmetry, so is $y \mapsto d(x, y)$ for each fixed x. Indeed, the mapping $d \colon E \times E \mapsto \mathbb{R}_+$ is continuous (which implies the preceding two statements). Hint: let (x_n) and (y_n) be sequences in E converging to x and y, and show that $d(x_n, y_n) \to d(x, y)$.

3.16 *Continuity of pairs.* Let $f \colon E \mapsto E'$ and $g \colon E \mapsto E''$ be continuous. Define $h \colon E \mapsto E' \times E''$ by letting $h(x)$ be the pair $(f(x), g(x))$. Show that h is continuous.

3.17 *Closed sets.* If $T \colon E \mapsto E'$ is continuous, then $T^{-1}B$ is a closed subset of E for every closed subset B of E'. Show. For $f \colon E \mapsto \mathbb{R}$ continuous, show that the sets $\{x \in E : f(x) \le b\}, \{x \in E : f(x) = b\}, \{x \in E : f(x) \ge b\}$ are closed in E.

3.18 *Indicators.* For $A \subset E$ let 1_A be the indicator of A, that is, $1_A(x) = 1$ if $x \in A$ and $1_A(x) = 0$ if $x \notin A$. Show that 1_A is continuous at all points x in E except for $x \in \partial A$.

3.19 *Left-continuity, right-continuity.* Let $f \colon \mathbb{R} \mapsto E'$. Order properties of the real line enable us to refine the notion of continuity as follows. The function f is said to be *right-continuous* at x in $\mathbb{R}$ provided that $f(x_n) \xrightarrow{d'} f(x)$ for every decreasing sequence $(x_n) \subset \mathbb{R}$ with limit x. Similarly, f is said to be *left-continuous* at x if $f(x_n) \xrightarrow{d'} f(x)$ for every increasing sequence (x_n) with limit x.

Show that f is continuous at x if and only if it is both right-continuous and left-continuous at x.

3.20 *Functional inverses.* Let $f: \mathbb{R}_+ \mapsto \mathbb{R}_+$ be a continuous and strictly increasing bijection. Let $f^{-1}(y)$ be that point x for which $f(x) = y$. Show that the function f^{-1} is continuous and strictly increasing.

3.21 *Legendre transforms.* A function $f: \mathbb{R} \mapsto \mathbb{R}$ is called *convex* if

$$f(px + qy) \le pf(x) + qf(y)$$

for all x, y in $\mathbb{R}$ and all p, q in $(0, 1)$ satisfying $p + q = 1$. The *Legendre transform* of a convex function f is the function $g: \mathbb{R} \mapsto \mathbb{R}$ defined by

$$g(y) = \max_x (xy - f(x)).$$

Show that g is convex and that

$$f(x) = \max_y (xy - g(y)).$$

State any extra "smoothness" assumptions you might need.

3.22 *Sections.* Let $f: E_1 \times E_2 \mapsto \mathbb{R}$ be continuous. Show that, for each y in E_2, the mapping $x \mapsto f(x, y)$ from E_1 into $\mathbb{R}$ is continuous. Similarly, $y \mapsto f(x, y)$ is continuous for each x. Unfortunately, the converse does not hold: it is possible to have $x \mapsto f(x, y)$ continuous for each y and $y \mapsto f(x, y)$ continuous for each x even though f is not continuous. Give an example of such a function.

B. Compactness and Uniform Continuity

As before, E, E', etc. are metric spaces with metrics d, d', etc. This section is on the effect of compactness on continuity.

3.23 THEOREM. Let $T: E \mapsto E'$ be continuous. If E is compact, then the range of T is a compact subset of E'.

PROOF. Let $D \subset E'$ be the range of T. Assuming that E is compact, we need to show that D is compact. Let $\{B_i : i \in I\}$ be a collection of open subsets of E' that covers D. Then, the continuity of T implies via Theorem 3.3 that the sets $A_i = T^{-1}B_i$,

$i \in I$, are open. Moreover, $\{A_i : i \in I\}$ covers E: if x is in E then Tx is in D, and hence, Tx is in B_i for some i, which implies that x is in the corresponding A_i. Now the compactness of E implies that there exists a finite subset J of I such that $\{A_i : i \in J\}$ covers E. Thus, if $x \in E$, then $x \in A_i$ for some i in J and therefore $Tx \in B_i$ for some i in J. That is, $\{B_i : i \in J\}$ covers D. So, D must be compact.   □

Recall that every compact set is closed and bounded. Thus, if $f : E \mapsto \mathbb{R}$ is continuous and E is compact, then the range of f is bounded and closed, which implies that f attains a maximum and a minimum, that is, there are x_0 and x_1, such that $f(x_0) \leq f(x) \leq f(x_1)$ for all x in E (see Exercise 2.81 to the effect that if a subset D of $\mathbb{R}$ is closed and bounded then inf A and sup A belong to D). We have thus shown the following result.

3.24 COROLLARY. Let E be compact and $f : E \mapsto \mathbb{R}$ continuous. Then, f is bounded and attains a maximum and a minimum.

The conclusion fails if E is not compact. For instance, $f(x) = x$ on $E = (0, 1)$ is bounded but has neither a maximum nor a minimum. Also, $f(x) = 1/x$ on $E = (0, 1)$ is not bounded and has neither a maximum nor a minimum.

Uniform Continuity

Recall the definition of continuity: a mapping $T : E \mapsto E'$ is continuous provided that for every x in E and every $\epsilon > 0$ there is a number $\delta > 0$ (depending on x and ϵ) such that $d(x, y) < \delta$ implies $d'(Tx, Ty) < \epsilon$ for all y in E. The import of the following is to remove the dependence of δ on x.

3.25 DEFINITION. A mapping $T : E \mapsto E'$ is said to be *uniformly continuous* provided that for every $\epsilon > 0$ there is a number $\delta > 0$ such that

$$x, y \in E, \quad d(x, y) < \delta \quad \Rightarrow \quad d'(Tx, Ty) < \epsilon.$$

Obviously, every uniformly continuous function is continuous. The converse is false. For example, the function $f : (0, 1) \mapsto \mathbb{R}$ defined by $f(x) = 1/x$ is continuous but not uniformly so. The failure here is not due to the unboundedness of f. For instance, the function $f : (0, 1) \mapsto [-1, 1]$ defined by $f(x) = \sin(1/x)$ is continuous but not uniformly so. The mappings of Examples 3.6, 3.7, 3.10, and 3.11 are uniformly continuous. In fact, they are all special cases of Example 3.10 on Lipschitz continuity. Being Lipschitz almost encapsulates the notion of uniform continuity.

3.26 PROPOSITION. Let $T\colon E \mapsto E'$ be Lipschitz continuous. Then T is uniformly continuous.

PROOF. Fix $\epsilon > 0$ and choose $\delta = \epsilon/K$. This δ works and is independent of x.   □

See Exercise 3.34 for an "almost converse" to this result. The following shows the important role of compactness in relation to uniform continuity.

3.27 THEOREM. Let $T\colon E \mapsto E'$ be continuous. If E is compact, then T is uniformly continuous.

PROOF. Fix $\epsilon > 0$. We search for $\delta > 0$ that will fulfill the condition for uniform continuity. Since T is continuous, for each x in E there is $\delta(x) > 0$ such that

3.28 $$d(x, y) < \delta(x) \Rightarrow d'(Tx, Ty) < \epsilon/2.$$

The collection of open balls $B(x, \delta(x)/2)$, $x \in E$, covers E. Since E is compact, there must exist a finite number of them, say those corresponding to $x_1, \ldots, x_n$, that cover E. Define

$$\delta = \frac{1}{2} \min\{\delta(x_1), \ldots, \delta(x_n)\}.$$

Then, $\delta > 0$ and it remains to show that this δ works. Let x, y in E be arbitrary and suppose that $d(x, y) < \delta$. By the way the $x_1, \ldots, x_n$ are chosen, there is an i such that x is in $B(x_i, \delta(x_i)/2)$, that is,

$$d(x, x_i) < \frac{1}{2}\delta(x_i).$$

Moreover, for the same i,

$$d(y, x_i) \le d(y, x) + d(x, x_i) \le \delta + \frac{1}{2}\delta(x_i) \le \delta(x_i).$$

Thus, $d(x, x_i) < \delta(x_i)$ and $d(y, x_i) < \delta(x_i)$, which by 3.28 imply that

$$d'(Tx, Tx_i) < \epsilon/2, \text{ and } d'(Ty, Tx_i) < \epsilon/2.$$

Thus, $d'(Tx, Ty) < \epsilon$ by the triangle inequality.   □

Exercises

3.29 *Metrics.* Show that, for fixed x_0 in E, the function $x \mapsto d(x, x_0)$ from E into $\mathbb{R}_+$ is uniformly continuous.

3.30 *Compositions.* Let $T \colon E \mapsto E'$ and $S \colon E' \mapsto E''$ be uniformly continuous. Show that, then, $S \circ T \colon E \mapsto E''$ is uniformly continuous.

3.31 *Homeomorphisms.* Recall that for a bijection $f \colon E \mapsto E'$ we define the functional inverse f^{-1} by setting $f^{-1}(y) = x$ if and only if $f(x) = y$. A *homeomorphism* from E onto E' is a bijection that is continuous and whose functional inverse is also continuous. Incidentally, two spaces E and E' are said to be *homeomorphic* if there exists a homeomorphism from one to the other. Compactness helps in checking for homeomorphisms. Show that if $f \colon E \mapsto E'$ is a continuous bijection and E is compact, then f is a homeomorphism.

3.32 *Extensions.* Let D be dense in E (see Exercise 2.46 for the definition). Note that this means that every point of $E \setminus D$ is a cluster point of D. Suppose that $f \colon D \mapsto \mathbb{R}$ is uniformly continuous. Show that, then, there exists a unique continuous function $\bar{f} \colon E \mapsto \mathbb{R}$ such that $\bar{f}(x) = f(x)$ for all x in D. Then, $\bar{f}$ is called the *continuous extension* of f onto E.

3.33 *Cantor function.* Let $E = [0, 1]$, and C be the Cantor set, and $D = E \setminus C$; see Example 2.40. Note that D is dense in E, since C contains no open intervals.

Show that the function f constructed in Example 2.40 of Chap. 2 is a uniformly continuous function from D into $[0, 1]$. By the preceding exercise, then, f has a continuous extension $\bar{f}$ onto $E = [0, 1]$. In fact, $\bar{f}$ is uniformly continuous (why?).

The function $\bar{f}$ is called the *Cantor function.* It is increasing and continuous. Its derivative exists at every x in D and is equal to 0. So, although $\bar{f}$ increases from 0 to 1 in a continuous fashion, all its increase is on the set C, and C has "length" 0.

3.34 *Lipschitz continuity.* A mapping $T \colon \mathbb{R}^n \mapsto \mathbb{R}$ is uniformly continuous if and only if for every $\epsilon > 0$ there exists K_ϵ such that

$$|Tx - Ty| \le K_\epsilon \cdot \|x - y\| + \epsilon$$

for all x and y in $\mathbb{R}^n$. Prove this.

Hints: (a) The "if" part is easy. Choose $\delta = \epsilon/2K_{\epsilon/2}$.

(b) For the "only if" part: fix $\epsilon > 0$ and x and y; choose a chain of points $x = x_0, x_1, x_2, \ldots, x_m = y$ with distances $\|x_i - x_{i+1}\| < \delta$; ask, how many such points do we need, and note that

$$|Tx - Ty| \leq \sum_1^m |Tx_i - Tx_{i+1}| \leq n\epsilon;$$

figure out the m needed and then what K_ϵ should be.

C. Sequences of Functions

Let E and E' be metric spaces with respective metrics d and d'. Let (T_n) be a sequence of mappings from E into E'.

3.35 DEFINITION. The sequence (T_n) is said to *converge pointwise* to a mapping $T: E \mapsto E'$ provided that the sequence $(T_n x)$ converges to Tx in E' for each point x in E.

In other words, for each x in E, we must have

3.36 $$\lim_n d'(T_n x, Tx) = 0,$$

that is, for every $\epsilon > 0$ there must be an integer $n_{\epsilon,x}$ such that $d'(T_n x, Tx) < \epsilon$ for all $n \geq n_{\epsilon,x}$. If $n_{\epsilon,x}$ can be chosen to be free of x, we obtain the following stronger concept of convergence.

3.37 DEFINITION. The sequence (T_n) is said to *converge uniformly* to a mapping T provided that

$$\lim_n \sup_{x \in E} d'(T_n x, Tx) = 0.$$

Obviously, uniform convergence of (T_n) implies pointwise convergence (and the limit T is the same). That the converse is generally false can be seen from Figs. 3.1 and 3.2: here the functions $f_n: \mathbb{R}_+ \mapsto [0, 1]$ converge pointwise, but not uniformly (Fig. 3.3).

Cauchy Criterion

As with sequences of points, it is important to have a criterion for the uniform convergence of (T_n) expressed in terms of the T_n themselves. The following Cauchy criterion does this.

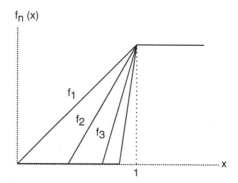

FIGURE 3.1. Here (f_n) converges to f, where $f(x) = 0$ for $x < 1$ and $f(x) = 1$ for $x \geq 1$. Convergence is pointwise but not uniform.

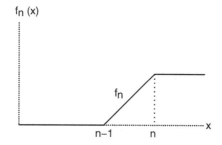

FIGURE 3.2. These f_n converge to $f = 0$ pointwise, but not uniformly.

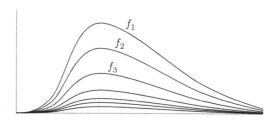

FIGURE 3.3. These f_n converge to 0 uniformly (and hence pointwise).

3.38 THEOREM. Suppose that E' is complete. Then, (T_n) is uniformly convergent if and only if for every $\epsilon > 0$ there is an integer n_ϵ with

3.39 $$\sup_x d'(T_n x, T_m x) < \epsilon \quad \text{for all } m > n \geq n_\epsilon.$$

PROOF. Suppose that (T_n) converges uniformly, say, to T. Then, for every $\epsilon > 0$, there is an integer n_ϵ such that $d'(T_n x, Tx) < \epsilon/2$ for all $n \geq n_\epsilon$. Thus, for $m, n \geq n_\epsilon$,

$$d'(T_n x, T_m x) \leq d'(T_n x, Tx) + d'(Tx, T_m x) < \epsilon/2 + \epsilon/2 = \epsilon$$

for all x. So, (T_n) is Cauchy (for every $\epsilon > 0$ there is n_ϵ such that 3.39 holds).

Let (T_n) be Cauchy. Then, in particular, for each x in E, the sequence $(T_n x)$ in E' is Cauchy. Since E' is complete, this implies that $(T_n x)$ converges to some point of E', call it Tx. This defines a mapping $T : E \mapsto E'$. We want to show that (T_n) converges to T uniformly. Since (T_n) is Cauchy, for every $\epsilon > 0$ there is an integer n_ϵ such that

$$d'(T_n x, T_m x) < \epsilon \quad \text{for all } m, n \geq n_\epsilon$$

for all x. Now, let $m \to \infty$; then, $(T_m x)$ converges to Tx and the continuity of $y \mapsto d'(T_n x, y)$ implies that $d'(T_n x, T_m x) \to d'(T_n x, Tx)$. Thus, as we needed to show, for $\epsilon > 0$ there is an integer n_ϵ with

$$d'(T_n x, Tx) < \epsilon \quad \text{for all } n \geq n_\epsilon \text{ and all } x \in E.$$

$\square$

Continuity of Limit Functions

As can be seen from Fig. 3.1, the pointwise limit of a sequence of continuous functions is not necessarily continuous. In fact, the primary use of uniform convergence is to ensure the continuity of the limit function.

3.40 THEOREM. Suppose that each T_n is continuous and (T_n) converges to T uniformly. Then, T is continuous.

PROOF. Fix x in E. Note that for all n and y

$$d'(Tx, Ty) \leq d'(Tx, T_n x) + d'(T_n x, T_n y) + d'(T_n y, Ty).$$

Given $\epsilon > 0$, there is an integer n_ϵ such that the first and third terms on the right side are less than $\epsilon/3$ each for $n = n_\epsilon$; this comes from the uniform convergence of (T_n) to T. Moreover, the continuity of T_{n_ϵ} at the point x implies the existence of $\delta = \delta_{\epsilon, x}$ such that the second term on the right with $n = n_\epsilon$ is less than $\epsilon/3$ for all y in $B(x, \delta)$. Hence, for every $\epsilon > 0$ there is a $\delta = \delta_{\epsilon, x}$ such that $d(x, y) < \delta$ implies that $d'(Tx, Ty) < \epsilon$ for all y; that is, T is continuous at x. $\square$

Exercises

3.41 Let $0 \leq a < b < 1$. Let $f_n : [a, b] \mapsto \mathbb{R}_+$ be defined by $f_n(x) = x^n$. Show that (f_n) converges uniformly to $f = 0$.

3.42 Let $T_n \colon [0,1] \mapsto [0,1]$ be defined by $T_n x = x^n(1-x)$. Show that (T_n) is uniformly convergent.

3.43 Let $f \colon \mathbb{R} \mapsto \mathbb{R}$ be uniformly continuous. Define $f_n(x) = f(x + 1/n)$. Show that (f_n) converges uniformly to f.

3.44 Let (f_n) be defined as a sequence of functions from $\mathbb{R}_+$ into $\mathbb{R}_+$ by $f_1(x) = \sqrt{x}$, $f_2(x) = \sqrt{x + \sqrt{x}}$, $f_3(x) = \sqrt{x + \sqrt{x + \sqrt{x}}}$, ... Show that (f_n) is convergent and find the limit function.

D. Spaces of Continuous Functions

Throughout this section (E, d) will be a compact metric space, and all functions are from E into $\mathbb{R}$. On a first reading, the reader should take $E = [a, b]$, a closed interval. Our aim is to illustrate the uses of the foregoing concepts in the analysis of the function space $C(E \mapsto \mathbb{R})$ of all continuous functions from E into $\mathbb{R}$. For brevity, we write C for $C(E \mapsto \mathbb{R})$.

The set C is a vector space: if f and g are in C then so is $af + bg$ for each a in $\mathbb{R}$ and b in $\mathbb{R}$. Moreover, various arithmetic operations are well-defined on C: $f + g$, $f - g$, $f \cdot g$, and f/g all belong to C if f and g are in C, except that in the case of f/g one must worry about $g(x) = 0$.

Although each point of C is a function, in many respects C is like a Euclidean space. We may, for instance, define a norm on C as follows. Let $f \in C$. Being a continuous function on a compact metric space, f is bounded and attains its maximum and minimum. It follows that

3.45
$$\|f\| = \max_{x \in E} |f(x)|$$

is a well-defined positive real number; it is called the *norm* of f. It is indeed a norm:

3.46
$$\|f\| \geq 0; \quad \|f\| = 0 \text{ if and only if } f = 0;$$

3.47
$$\|cf\| = |c| \cdot \|f\|;$$

3.48
$$\|f + g\| \leq \|f\| + \|g\|.$$

As with Euclidean spaces, we may use the norm above to define a metric on C. We define the distance between f and g to be

3.49
$$d(f, g) = \|f - g\|.$$

Convergence in C

  The following shows that convergence in the metric space C is equivalent to uniform convergence of functions on E.

3.50 THEOREM. A sequence (f_n) in C is convergent if and only if the sequence of functions $f_n : E \mapsto \mathbb{R}$ is uniformly convergent.

PROOF. In view of the definition of convergence for a sequence of points in a metric space and the definition of uniform convergence for a sequence of functions $f_n : E \mapsto \mathbb{R}$, the claim is simply that

$$\lim_n d(f_n, f) = 0 \quad \Leftrightarrow \quad \lim_n \sup_{x \in E} |f_n(x) - f(x)| = 0.$$

But this is obvious in view of 3.49 and 3.45.   □

  Conceptually, then, the somewhat complex concept of uniform convergence of a sequence of functions is equivalent to the simpler concept of convergence of a sequence in a metric space.

Lipschitz Continuous Functions

  A function f in C is said to be *Lipschitz continuous* if there exists a constant K such that

3.51 $|f(x) - f(y)| \le K \cdot d(x, y) \quad$ for all $x, y \in E$.

The constant K is called the *Lipschitz constant*. For example, if $E = [a, b]$, f is differentiable, and the derivative f' is bounded by K, then 3.51 holds with the same K, and f is Lipschitz continuous.

  Let B_K be the set of all f in C satisfying 3.51. Then, clearly, the set of all Lipschitz continuous functions is exactly the union of the B_K's over all K in $\mathbb{R}_+$. The next theorem shows that each B_K is closed. Unfortunately, the union $\bigcup_K B_K$ is not closed, as can be seen from the sequence of functions shown in Fig. 3.4. In fact, the closure of this union is precisely C, that is, every f in C is the limit of a sequence of Lipschitz continuous functions; we leave this as an exercise to prove.

3.52 PROPOSITION. Each B_K is a closed subset of C.

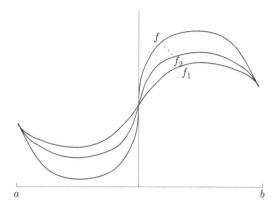

FIGURE 3.4. A sequence of Lipschitz continuous functions converging to a continuous function that is not Lipschitz.

PROOF. We use the characterization given in Theorem 2.55. Let (f_n) in B_K converge to the point f in $\mathcal{C}$. We need to show that f is in B_K. Now, for arbitrary x and y in E,

$$\begin{aligned} |f(x) - f(y)| &\leq |f(x) - f_n(x)| + |f_n(x) - f_n(y)| + |f_n(y) - f(y)| \\ &\leq \|f - f_n\| + Kd(x,y) + \|f_n - f\| \end{aligned}$$

for all n. Since $\|f_n - f\| \to 0$, this shows that f satisfies 3.51. $\qquad\square$

Completeness

The space $\mathcal{C}$ is not bounded. Therefore it cannot be compact. But, at least, it is complete.

3.53 THEOREM. The space $\mathcal{C}$ is complete.

PROOF. Let $(f_n) \subset \mathcal{C}$ be Cauchy, that is, for every $\epsilon > 0$ there is an integer n_ϵ such that $\|f_n - f_m\| \leq \epsilon$ for all $m > n \geq n_\epsilon$. This is equivalent to the condition 3.39 (here $E' = \mathbb{R}$ which is complete). Thus, by Theorem 3.38, (f_n) is uniformly convergent as a sequence of functions on E. But, by Theorem 3.50, uniform convergence is equivalent to convergence in $\mathcal{C}$. So, (f_n) is convergent in $\mathcal{C}$. $\qquad\square$

Functionals

Since $\mathcal{C} = C(E \mapsto \mathbb{R})$ is a metric space, we may speak of functions defined on $\mathcal{C}$ as we speak of functions defined on E. For linguistic clarity, a function from $\mathcal{C}$ into $\mathbb{R}$ is called a *functional*. Here are some examples of functionals:

3.54 $$M \colon f \mapsto M(f) = \max_{x \in E} f(x),$$

3.55 $$P_x: f \mapsto P_x(f) = f(x), \quad x \in E \text{ fixed,}$$

3.56 $$F: f \mapsto \varphi(f(x_1), \ldots, f(x_k)),$$

where $k \geq 1$ is a fixed integer, $\varphi: \mathbb{R}^k \mapsto \mathbb{R}$ is fixed, and $x_1, \ldots, x_k$ are fixed in E. Here are some further examples in the particular case where $E = [a, b]$:

3.57 $$L(f) = \int_a^b f(x)\, dx,$$

3.58 $$L_\varphi(f) = \int_a^b \varphi(x) f(x)\, dx,$$

where $\varphi \in \mathcal{C}$ is some fixed function.

The functional M is uniformly continuous; in fact, it is Lipschitz continuous with Lipschitz constant $K = 1$:

$$
\begin{aligned}
|M(f) - M(g)| &= \left| \max_x f(x) - \max_x g(x) \right| \\
&\leq \max_x |f(x) - g(x)| = \|f - g\| = d(f, g).
\end{aligned}
$$

Even easier is the Lipschitz continuity of the coordinate mapping P_x for fixed x:

$$|P_x(f) - P_x(g)| = |f(x) - g(x)| \leq \|f - g\|.$$

Assuming that the function $\varphi: \mathbb{R}^k \mapsto \mathbb{R}$ is continuous, the functional F is continuous: if $\|f_n - f\| \to 0$, then the sequence of points $(f_n(x_1), \ldots, f_n(x_k))$ in $\mathbb{R}^k$ converges to the point $(f(x_1), \ldots, f(x_k))$ of $\mathbb{R}^k$ as $n \to \infty$, and the continuity of φ implies that $F(f_n) \to F(f)$.

The functional L is a linear transformation from $\mathcal{C}$ into $\mathbb{R}$. It is uniformly continuous; in fact, it is Lipschitz continuous with Lipschitz constant $K = b - a$. So is L_φ, with Lipschitz constant $K = \int_a^b |\varphi(x)|\, dx$.

Exercises

3.59 If f and g are continuous functions on a compact metric space, show that

$$\left| \max_x f(x) - \max_x g(x) \right| \leq \max_x |f(x) - g(x)|.$$

3.60 *Differentiable functions.* For fixed K in $\mathbb{R}_+$, let A_K denote the set of all differentiable functions f whose derivatives f' are continuous and bounded by K. The set A_K is not closed, which can be seen from Fig. 3.5 where $(f_n) \subset A_K$, (f_n) converges to f in $\mathcal{C}$, but f is not in A_K. In fact, the closure of A_K is precisely B_K. Show this.

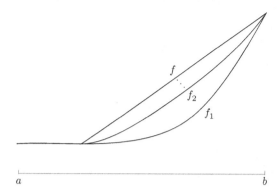

FIGURE 3.5. A sequence of differentiable functions whose derivatives are bounded but whose limit is not differentiable.

3.61 *Closure of $\bigcup_K B_K$ is $\mathcal{C}$.* For each f in $\mathcal{C}$ there exists a sequence (f_n) of Lipschitz continuous functions whose limit is f. Prove this.

CHAPTER 4

Differential and Integral Equations

The aim of this chapter is to discuss several applications of metric space ideas to some classical problems of engineering analysis.

We shall start with one theorem, the fixed point theorem for contractions on a metric space, and show how various problems can be beaten into submission with it.

A. Contraction Mappings

The aim of this section is to prepare the stage for some applications for differential and integral equations encountered frequently in engineering. Throughout, E is a metric space with some metric d.

We shall use the term *transformation on E* to mean a mapping from E into E. If T is a transformation on E, then the image Tx of x is a point in E, which allows us to apply T to the point Tx; the image of Tx is $T(Tx)$, for which we write T_2x. In other words, we are writing T_2 for $T \circ T$. Further iterates are defined by

$$4.1 \qquad T_{n+1}x = T(T_nx), \quad x \in E, n \in \mathbb{N},$$

with $T_0x = x$. So, T_0 is the identity, $T_1 = T$, etc.

Given a point x in E, the sequence (x_n) obtained by putting $x_n = T_nx$, $n \in \mathbb{N}$, is called the *orbit* of x. One should think of x_n as the position at time n of a particle that starts at $x_0 = x$ and moves successively to $x_1 = Tx_0$, $x_2 = Tx_1$, ...; see Fig. 4.1.

Contractions

Let (E, d) be a metric space. A transformation T on E is said to be a *contraction* if it is Lipschitz continuous with some Lipschitz constant $\alpha < 1$. In other words, T is a contraction of E if there exists a constant α in $[0, 1)$ such that

$$4.2 \qquad d(Tx, Ty) \leq \alpha \cdot d(x, y) \quad \text{for all } x, y \in E.$$

Then, for the iterates of T we obtain

$$4.3 \qquad d(T_nx, T_ny) \leq \alpha^n \cdot d(x, y) \quad \text{for all } x, y \in E, \ n \in \mathbb{N}.$$

E. Çınlar and R.J. Vanderbei, *Real and Convex Analysis*, Undergraduate Texts in Mathematics, DOI 10.1007/978-1-4614-5257-7_4, © Springer Science+Business Media New York 2013

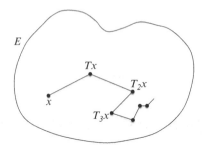

FIGURE 4.1. The orbit of x under the map T.

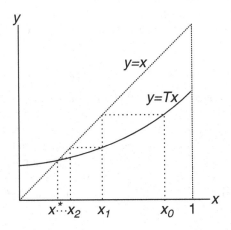

FIGURE 4.2. A contraction T on $[0, 1]$ and its fixed point x^*.

Thinking of the orbits of x and y, we see that the two particles starting at x and y are approaching each other geometrically quickly.

Fixed Point Theorem

Let T be a transformation. A point x is said to be a *fixed point* of T if $Tx = x$. Then, the orbit of x is stationary; that is, $T_n x = x$ for all n. Figure 4.2 shows a transformation T on $E = [0, 1]$; there, x^* is the unique fixed point of T, and the orbit $(T_n x_0)$ of x_0 converges to the fixed point x^*.

The following theorem shows that every contraction of a complete metric space has a unique fixed point. Its proof shows how to obtain the fixed point by the method of successive approximations.

4.4 THEOREM. Suppose that E is complete. Let T be a contraction on E. Then, T has a unique fixed point, and, for each point x_0 in E, the orbit of x_0 converges to that fixed point.

PROOF. Fix x_0 in E, let (x_n) be its orbit, and put $c = d(x_0, x_1)$ for simplicity. Observe from 4.1 and 4.3 that

$$d(x_n, x_{n+1}) = d(T_n x_0, T_n x_1) \leq \alpha^n d(x_0, x_1) = c\alpha^n$$

for every n. Thus, for $n < m$, using the triangle inequality,

$$
\begin{aligned}
d(x_n, x_m) &\leq d(x_n, x_{n+1}) + d(x_{n+1}, x_{n+2}) + \cdots + d(x_{m-1}, x_m) \\
&\leq c\alpha^n + c\alpha^{n+1} + \cdots + c\alpha^{m-1} \leq c\alpha^n(1 + \alpha + \alpha^2 + \cdots) = c\frac{\alpha^n}{1 - \alpha}.
\end{aligned}
$$

Since $\alpha < 1$, the right side goes to 0 as $n \to \infty$. Hence, the sequence (x_n) is Cauchy and must converge to some point x in E in view of the hypothesis that E is complete. Then, by the continuity of T,

$$Tx = T(\lim x_n) = \lim T x_n = \lim x_{n+1} = x,$$

that is, x is a fixed point.

To complete the proof, we now show that the fixed point is unique. To this end, let x and y be fixed points. Then, since $Tx = x$ and $Ty = y$, we have via 4.2 that

$$d(x, y) \leq \alpha d(x, y).$$

Since $\alpha < 1$, this is possible only if $d(x, y) = 0$, that is, $x = y$. $\square$

The preceding theorem can be used to prove existence and uniqueness of solutions to a wide variety of equations. Besides showing that $Tx = x$ has a solution, the proof gives a practical method for arriving at it. Indeed, start from an arbitrary point x_0 and successively compute $x_1 = Tx$, $x_2 = Tx_1$, $x_3 = Tx_2$, ... The x_n get close to x geometrically fast: since $x_n = T_n x_0$ and $x = T_n x$,

$$d(x_n, x) = d(T_n x_0, T_n x) \leq \alpha^n d(x_0, x).$$

Exercises

4.5 For the transformation $T \colon [0, 1] \mapsto [0, 1]$ shown in Fig. 4.3, find the orbit of the point x_0 indicated.

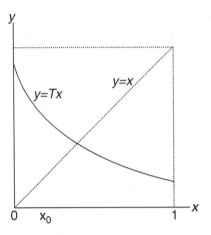

FIGURE 4.3. Exercise 4.5.

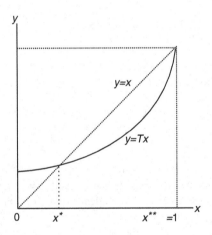

FIGURE 4.4. Exercise 4.6.

4.6 For the transformation $T\colon [0,1] \mapsto [0,1]$ given by $Tx = 0.3 + 0.2x + 0.5x^3$, Fig. 4.4 shows that there are exactly two fixed points. Find them. Show that, for arbitrary $x_0 \neq 1$, the orbit of x_0 converges to the smaller fixed point x^*.

4.7 *Branching processes.* In a chain reaction, each particle gives rise to a random number of new particles. Each of these new particles acts independently and produces random numbers of newer particles. And this continues indefinitely. Let p_k be the

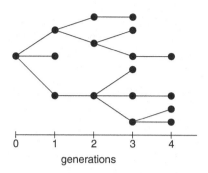

generations

FIGURE 4.5. Exercise 4.7.

probability that a particle produces k particles; here $p_0, p_1, p_2, \ldots$ are positive numbers with $\sum p_k = 1$. Starting with one particle, we now consider the probability that the chain reaction fizzles out, that is, the population of particles becomes extinct. Let x_n be the probability that the nth generation is extinct already (Fig. 4.5). Note that the $(n+1)$th generation consists of particles that are nth-generation offspring of the individuals of the first generation. In order for the population to be extinct at or before the $(n+1)$th generation, populations initiated by the particles of the first generation must all become extinct. Thus,

$$x_{n+1} = \sum_{k=0}^{\infty} p_k (x_n)^k.$$

In other words, $x_{n+1} = Tx_n$ where $T \colon [0,1] \mapsto [0,1]$ is defined by

$$Tx = \sum_{k=0}^{\infty} p_k x^k, \quad x \in [0,1].$$

Now, the probability x^* of eventual extinction for the population is the limit of x_n, and thus satisfies

$$x^* = Tx^*.$$

(a) Show that $x_1 = p_0$. Show that the sequence (x_n) increases to the extinction probability x^*.

(b) Assume that $p_0 > 0$. If $p_0 + p_1 = 1$ (so that $p_2 = p_3 = \cdots = 0$) show that $x^* = 1$.

(c) Show that the mapping $x \mapsto Tx$ is increasing and convex.

(d) Let $a = \sum_{k=1}^{\infty} p_k k$, that is, a is the expected number of particles produced by one particle. Show that if $a \leq 1$, then $x = Tx$ has only one solution and the fixed point is $x^* = 1$.

(e) Suppose that $a > 1$. Then, show that $x = Tx$ has exactly two solutions. One solution is 1, the other is the extinction probability x^*. Show this by examining the graph of T and using (a).

4.8 Let $T\colon [0,1] \mapsto [0,1]$ be defined by

$$Tx = 4x(1-x).$$

Show that T has exactly two fixed points. Compute them. Give an example of an orbit that converges to the fixed point $x^* = 0$. Note the highly chaotic nature of the orbits.

4.9 Let $T\colon [0,1] \mapsto [0,1]$ be defined by $Tx = 2x \pmod 1$, that is, $Tx = 2x$ if $2x < 1$ and $Tx = 2x - 1$ if $2x \geq 1$. The only fixed point is $x^* = 0$.

Incidentally, if $x = 0.\omega_1\omega_2\omega_3 \cdots$ is the binary representation of x then $Tx = 0.\omega_2\omega_3\omega_4 \cdots$ and $T_2x = 0.\omega_3\omega_4\omega_5 \cdots$, etc. Note the highly chaotic nature of the orbits by plotting $(T_n x)$.

4.10 Let $T\colon \mathbb{R}^n \mapsto \mathbb{R}^n$ be a linear transformation, say $Tx = Ax$ where A is some $n \times n$ matrix. Give a condition on A that guarantees T to be a contraction (with the Euclidean metric on $\mathbb{R}^n$).

4.11 Let $Tx = Ax + b$ where A is an $n \times n$ matrix and b is a fixed vector in $\mathbb{R}^n$. Consider $E = \mathbb{R}^n$ with the weighted Manhattan metric $d(x,y) = \sum_{i=1}^{n} w_i \cdot |x_i - y_i|$ where the weights $w_1, \ldots, w_n$ are strictly positive. Show that, to establish that T is a contraction of this metric space E, it is sufficient to have

$$\sum_{i=1}^{n} w_i |a_{ij}| < w_j, \quad j = 1, \ldots, n.$$

B. Systems of Linear Equations

In this section we discuss the use of the fixed point theorem in solving systems of linear equations. As a by-product, we get a chance to discuss the importance of choosing the right metric for a particular application.

Let $E = \mathbb{R}^n$; we do not specify the metric just yet. Fix b in $\mathbb{R}^n$ and consider the system of linear equations

4.12
$$x_i = \sum_{j=1}^{n} a_{ij} x_j + b_i, \quad i = 1, \ldots, n,$$

where the a_{ij} are real numbers. Writing A for the $n \times n$ matrix of elements a_{ij}, the system 4.12 is equivalent to

4.13
$$x = Ax + b.$$

In other words, the problem is to find the fixed point of the transformation $T \colon \mathbb{R}^n \mapsto \mathbb{R}^n$ defined by

4.14
$$Tx = Ax + b.$$

If T is a contraction, then we can use Theorem 4.4 and obtain the unique solution of $Tx = x$ by the method of successive approximations.

The conditions under which T is a contraction depend on the choice of metric on $E = \mathbb{R}^n$. We discuss three cases.

Maximum Norm

Suppose that d is the metric associated with the maximum norm:

$$d(x, y) = \max_{1 \le i \le n} |x_i - y_i|.$$

Then, since $Tx - Ty = Ax - Ay = A(x - y)$,

$$
\begin{aligned}
d(Tx, Ty) &= \max_i \left| \sum_{j=1}^{n} a_{ij}(x_j - y_j) \right| \\
&\le \max_i \sum_j |a_{ij}| \cdot |x_j - y_j| \\
&\le \max_i \sum_j |a_{ij}| \max_k |x_k - y_k| = \left(\max_i \sum_j |a_{ij}| \right) d(x, y).
\end{aligned}
$$

Thus, the contraction condition 4.2 is satisfied if

4.15
$$\alpha = \max_i \sum_j |a_{ij}| < 1.$$

Manhattan Metric

Suppose that d is the Manhattan metric:

$$d(x, y) = \sum_{i=1}^{n} |x_i - y_i|.$$

Then,

$$
\begin{aligned}
d(Tx, Ty) &= \sum_i \left| \sum_j a_{ij}(x_j - y_j) \right| \\
&\leq \sum_i \sum_j |a_{ij}| \cdot |x_j - y_j| \leq \left(\max_j \sum_i |a_{ij}| \right) d(x, y),
\end{aligned}
$$

and the contraction condition is satisfied if

4.16
$$
\alpha = \max_j \sum_i |a_{ij}| < 1.
$$

Euclidean Metric

Suppose that d is the ordinary Euclidean distance. Then,

$$
\begin{aligned}
d(Tx, Ty)^2 &= \sum_i \left(\sum_j a_{ij}(x_j - y_j) \right)^2 \\
&\leq \sum_i \left(\sum_j a_{ij}^2 \right) \left(\sum_j (x_j - y_j)^2 \right) = \left(\sum_i \sum_j a_{ij}^2 \right) d(x, y)^2,
\end{aligned}
$$

where we used Schwarz's inequality at the second step. Thus, the contraction condition 4.2 is satisfied if

4.17
$$
\alpha = \sum_i \sum_j a_{ij}^2 < 1.
$$

Conclusion

Under each of the metrics discussed, $\mathbb{R}^n$ is a complete metric space. Hence, if at least one of the conditions 4.15–4.17 holds, Theorem 4.4 applies to show that there exists a unique solution to 4.12. The solution x^* is a fixed point of T, and it can be obtained as the limit of the sequence of approximations

4.18
$$
x, \; Tx, \; T_2x, \; \ldots
$$

starting with an arbitrary initial point x. However, none of the conditions 4.15–4.17 is necessary; it is easy to give examples of A where one condition holds but not the others.

C. Integral Equations

The most interesting applications of fixed point theorems arise when the underlying metric space is a function space. Here we discuss the existence and uniqueness of solutions to Fredholm and Volterra equations.

Fredholm Equation

A *Fredholm equation* (of the second kind) is an integral equation of the form

4.19
$$f(x) = \varphi(x) + \lambda \int_a^b K(x,y) f(y)\, dy.$$

Here, the functions $\varphi\colon [a,b] \mapsto \mathbb{R}$ and $K\colon [a,b] \times [a,b] \mapsto \mathbb{R}$ are given, λ is an arbitrary fixed real number, and $f\colon [a,b] \mapsto \mathbb{R}$ is the unknown function. The function K is called the *kernel* of the equation. The equation is said to be *homogeneous* if $\varphi = 0$, and *nonhomogeneous* otherwise.

The Fredholm equation is the continuous version of the system of linear Eqs. 4.12. To see this, suppose that the interval is discretized and is replaced by $n+1$ equidistant points $a = x_0 < x_1 < \cdots < x_n = b$. Then, writing $y_i = f(x_i)$ and $b_i = \varphi(x_i)$ and $a_{ij} = \lambda K(x_i, x_j)/n$, we see that 4.19 becomes

$$y_i = b_i + \sum_j a_{ij} y_j.$$

Whether this discretization is appropriate is a different matter; but it serves to visualize 4.19 as a generalization of 4.13.

Let $\mathcal{C} = C([a,b] \mapsto \mathbb{R})$, the collection of all continuous functions f from $[a,b]$ into $\mathbb{R}$, and let the metric on $\mathcal{C}$ be defined through the supremum norm:

4.20
$$d(f,g) = \|f - g\| = \sup_{a \le x \le b} |f(x) - g(x)|.$$

With this metric, $\mathcal{C}$ is a complete metric space (see Theorem 3.53).

Suppose that K is continuous on the square $[a,b] \times [a,b]$ and that φ is continuous on $[a,b]$. Then, the function Tf defined by

4.21
$$Tf(x) = \varphi(x) + \lambda \int_a^b K(x,y) f(y)\, dy$$

is continuous on $[a,b]$ for each continuous function f on $[a,b]$. In other words, the mapping $f \mapsto Tf$ is a transformation on $\mathcal{C}$. Now, the Fredholm Eq. 4.19 becomes

4.22
$$f = Tf,$$

and thus, solving 4.19 is equivalent to finding the fixed points of the transformation T on $\mathcal{C}$.

To this end, in order to apply the fixed point theorem (Theorem 4.4), all we need to show is that T is a contraction (recall that $\mathcal{C}$ is complete). The following shows that T is indeed so if the parameter λ is small enough.

4.23 THEOREM. Suppose that φ and K are continuous. Then there exists $\lambda_0 > 0$ such that Eq. 4.19 has a unique solution f for each λ in $(-\lambda_0, \lambda_0)$. Moreover, the solution f is continuous.

PROOF. Since K is continuous on the square $[a, b] \times [a, b]$, it is bounded there (continuous functions on compact spaces are bounded). So, there is a constant $c > 0$ such that $|K(x, y)| \le c$ for all x, y. Thus,

$$\begin{aligned}\|Tf - Tg\| &= \max_x \left| \lambda \int_a^b K(x, y)(f(y) - g(y))\, dy \right| \\ &\le |\lambda| \cdot c \cdot (b - a) \max_y |f(y) - g(y)| = |\lambda| \cdot c \cdot (b - a) \cdot \|f - g\|.\end{aligned}$$

Choose $\lambda_0 = 1/(c \cdot (b - a))$. Then, for each λ in $(-\lambda_0, \lambda_0)$, the preceding shows that T is a contraction on $\mathcal{C}$. By Theorem 4.4, consequently, there is a unique fixed point f in $\mathcal{C}$ of the transformation T. □

4.24 EXAMPLE. Suppose that $K(x, y) = xy$ on $[0, 1] \times [0, 1]$. Let $\varphi \in \mathcal{C}$ and consider the Fredholm equation

4.25
$$f(x) = \varphi(x) + \lambda \int_0^1 xy f(y)\, dy.$$

The proof of Theorem 4.23 shows that, for $|\lambda| < 1$, there is a unique solution f. And the solution is the limit of the sequence

$$f_0 = \varphi, \quad f_1 = Tf_0, \quad f_2 = Tf_1, \quad f_3 = Tf_2, \quad \dots$$

where, in general,

$$Tf(x) = \varphi(x) + \lambda x \int_0^1 y f(y)\, dy, \quad x \in [0, 1].$$

Now, we start computing. Defining $a = \int_0^1 y\varphi(y)\,dy$, we have

$$
\begin{aligned}
f_0(x) &= \varphi(x), \\
f_1(x) &= Tf_0(x) &&= \varphi(x) + \lambda x \int_0^1 y\varphi(y)\,dy \\
&&&= \varphi(x) + a\lambda x, \\
f_2(x) &= Tf_1(x) &&= \varphi(x) + \lambda x \int_0^1 y(\varphi(y) + a\lambda y)\,dy \\
&&&= \varphi(x) + a\lambda x + a\tfrac{\lambda^2}{3}x, \\
f_3(x) &= Tf_2(x) &&= \varphi(x) + \lambda x \int_0^1 y(\varphi(y) + a\lambda y + a\tfrac{\lambda^2}{3}y)\,dy \\
&&&= \varphi(x) + a\lambda x + a\tfrac{\lambda^2}{3}x + a\tfrac{\lambda^3}{9}x, \\
&\quad\vdots \\
f_n(x) &= Tf_{n-1}(x) &&= \varphi(x) + a\lambda x \left(1 + \tfrac{\lambda}{3} + \left(\tfrac{\lambda}{3}\right)^2 + \cdots + \left(\tfrac{\lambda}{3}\right)^{n-1}\right).
\end{aligned}
$$

It is clear from this that a fixed point f exists for every λ in $(-3, 3)$, and the solution to 4.25 is

4.26
$$
f(x) = \lim_n f_n(x) = \frac{3a\lambda}{3 - \lambda}x + \varphi(x).
$$

Going back to 4.25, the special form of the kernel K suggests a quicker method. Indeed, let

$$
c = \int_0^1 yf(y)\,dy.
$$

Then, using 4.25 in the form

$$
f(x) = \varphi(x) + \lambda x c,
$$

we get

$$
c = \int_0^1 xf(x)\,dx = \int_0^1 x\varphi(x)\,dx + \int_0^1 x\lambda x c\,dx = a + \frac{\lambda}{3}c.
$$

Solving this for c, we see that

$$
f(x) = \varphi(x) + \lambda x c = \varphi(x) + \frac{3a\lambda}{3 - \lambda}x
$$

as before provided that $\lambda \neq 3$. Note that this is the solution for arbitrary $\lambda \neq 3$. But the method of successive approximations works for $|\lambda| < 3$ only.

Studying the iterative method in the preceding example, we can get a theoretical understanding of the nature of solutions. To this end, we redo the computations of

$f_0 = \varphi$, $f_1 = Tf_0$, $f_2 = Tf_1$, ... once more, now with an arbitrary kernel K, and omitting the limits of integration. We get

$$
\begin{aligned}
f_0(x) &= \varphi(x), \\
f_1(x) &= Tf_0(x) &= \varphi(x) + \lambda \int K(x,y)\varphi(y)\,dy, \\
f_2(x) &= Tf_1(x) &= \varphi(x) + \lambda \int K(x,y)f_1(y)\,dy \\
&&= \varphi(x) + \lambda \int K(x,y)\left[\varphi(y) + \lambda \int K(y,z)\varphi(z)\,dz\right]dy \\
&&= \varphi(x) + \lambda \int K(x,y)\varphi(y)\,dy + \lambda^2 \int K_2(x,z)\varphi(z)\,dz
\end{aligned}
$$

where

$$
K_2(x,z) = \int K(x,y)K(y,z)\,dy.
$$

Continuing,

$$
\begin{aligned}
f_3(x) &= Tf_2(x) \\
&= \varphi(x) + \lambda \int K(x,y)\left[\varphi(y) + \lambda \int K(y,z)\varphi(z)\,dz\right. \\
&\qquad \left. + \lambda^2 \int K_2(y,z)\varphi(z)\,dz\right]dy \\
&= \varphi(x) + \lambda \int K(x,z)\varphi(z)\,dz \\
&\qquad + \lambda^2 \int K_2(x,z)\varphi(z)\,dz + \lambda^3 \int K_3(x,z)\varphi(z)\,dz
\end{aligned}
$$

where

$$
K_3(x,z) = \int K(x,y)K_2(y,z)\,dy.
$$

The pattern is now clear. We have

4.27
$$
f_n(x) = \varphi(x) + \sum_{i=1}^{n} \lambda^i \int_a^b K_i(x,y)\varphi(y)\,dy
$$

with $K_1 = K$, and K_2, K_3, ... defined recursively via

4.28
$$
K_{i+1}(x,y) = \int_a^b K(x,z)K_i(z,y)\,dz.
$$

Theorem 4.23 shows that when $|\lambda| < \lambda_0$, the sequence f_n converges to the fixed point f, where

4.29
$$
f(x) = \varphi(x) + \sum_{i=1}^{\infty} \lambda^i \int_a^b K_i(x,y)\varphi(y)\,dy.
$$

Since this is true for arbitrary φ, we can change the order of summation and integration. Thus, with

4.30
$$R_\lambda(x,y) = \sum_{i=1}^{\infty} \lambda^i K_i(x,y),$$

we have

4.31
$$f(x) = \varphi(x) + \int_a^b R_\lambda(x,y)\varphi(y)\,dy.$$

Although 4.28, 4.30, 4.31 together give an "explicit" solution to the Fredholm equation, this explicitness is only theoretical. For, computing R_λ is of the same order of difficulty as solving 4.19 (in fact, even harder).

On the other hand, if the kernel K is simple enough, analytic solutions might be possible. The following illustrates the computations for such a special case.

4.32 EXAMPLE. Suppose that

$$K(x,y) = \sum_{j=1}^{n} p_j(x)q_j(y), \quad x,y \in [a,b],$$

for some continuous functions $p_1, \ldots, p_n$ and $q_1, \ldots, q_n$ on $[a,b]$. For φ continuous on $[a,b]$, consider the Fredholm Eqs. 4.19. Now, if $f \in C$ satisfies 4.19, then

4.33
$$f(x) = \varphi(x) + \lambda \sum_{j=1}^{n} z_j p_j(x)$$

where

4.34
$$z_j = \int_a^b q_j(y)f(y)\,dy, \quad j = 1, \ldots, n.$$

In view of 4.33, then

$$
\begin{aligned}
z_i &= \int_a^b q_i(x)f(x)\,dx \\
&= \int_a^b q_i(x)\varphi(x)\,dx + \lambda \sum_{j=1}^{n} \left(\int_a^b q_i(x)p_j(x)\,dx \right) z_j.
\end{aligned}
$$

Thus, letting

4.35
$$c_i = \int_a^b q_i(x)\varphi(x)\,dx, \quad a_{ij} = \int_a^b q_i(x)p_j(x)\,dx,$$

we obtain

4.36
$$z_i = c_i + \lambda \sum_{j=1}^{n} a_{ij} z_j, \quad i = 1, 2, \ldots, n.$$

Note that the c_i and a_{ij} are known. If we can solve 4.36 for the z_i's, then 4.33 gives the solution f.

In vector–matrix notation, 4.36 becomes

$$z = c + \lambda A z,$$

whose solution is easy to discern. We can solve it for z (for arbitrary c) as long as $I - \lambda A$ is invertible, that is, as long as $1/\lambda$ is not an eigenvalue for A. Thus, we have a solution z for arbitrary b provided that $\lambda \in (-1/\lambda_0, 1/\lambda_0)$, where λ_0 is the modulus of the largest eigenvalue of A.

Volterra Equation

Let K be a continuous function on $[a, b] \times [a, b]$ and let φ be a continuous function on $[a, b]$. Consider the equation

4.37 $$f(x) = \varphi(x) + \lambda \int_a^x K(x, y) f(y)\, dy, \quad x \in [a, b].$$

It is called the *Volterra equation*. It differs from the Fredholm equation only slightly, and in form only. If we define

$$\hat{K}(x, y) = \begin{cases} K(x, y) & \text{if } y \le x, \\ 0 & \text{if } y > x, \end{cases}$$

then 4.37 becomes the Fredholm Eq. 4.19 with kernel $\hat{K}$. However, it is easier to attack 4.37 directly.

4.38 THEOREM. For each λ in $\mathbb{R}$, the Volterra Eq. 4.37 has a unique solution f that is continuous on $[a, b]$.

PROOF. Let $\mathcal{C} = C([a, b] \mapsto \mathbb{R})$, the set of all continuous functions from $[a, b]$ into $\mathbb{R}$, with the usual uniform metric $\|f - g\|$. Let c be the maximum of $|K(x, y)|$ over all x, y in $[a, b]$; this number is finite since K is continuous. Define the transformation $T \colon f \mapsto Tf$ on $\mathcal{C}$ by

$$Tf(x) = \varphi(x) + \lambda \int_a^x K(x, y) f(y)\, dy.$$

Now, for f and g in $\mathcal{C}$,

$$
\begin{aligned}
|Tf(x) - Tg(x)| &= \left| \lambda \int_a^x K(x, y)[f(y) - g(y)]\, dy \right| \\
&\le |\lambda|\, c\, (x - a)\, \|f - g\|, \quad x \in [a, b].
\end{aligned}
$$

We use this, next, to bound $T_2 f - T_2 g = T(Tf - Tg)$:

$$
\begin{aligned}
|T_2 f(x) - T_2 g(x)| &= \left| \lambda \int_a^x K(x,y)[Tf(y) - Tg(y)]\, dy \right| \\
&\leq |\lambda| \int_a^x |K(x,y)| \, |\lambda| \, c(y-a) \, \|f-g\| \, dy \\
&\leq |\lambda|^2 c^2 \int_a^x (y-a)\, dy \, \|f-g\| \\
&\leq \frac{|\lambda|^2 c^2 (x-a)^2}{2} \|f-g\|.
\end{aligned}
$$

Iterating in this manner, we see that

$$
|T_k f(x) - T_k g(x)| \leq \frac{|\lambda|^k c^k (x-a)^k}{k!} \|f-g\|
$$

for all x in $[a,b]$. Hence,

$$
\|T_k f - T_k g\| \leq \frac{[\,|\lambda|c(b-a)\,]^k}{k!} \|f-g\|.
$$

Recalling that $r^n/n!$ tends to 0 as $n \mapsto \infty$ for any r in $\mathbb{R}$, we conclude that there exists k such that T_k is a contraction: simply take k large enough to have $[\,|\lambda|c(b-a)\,]^k/k! < 1$. Finally, the existence and uniqueness of f in $\mathcal{C}$ satisfying $f = Tf$ follows from the next theorem. Obviously, if $f = Tf$, then f solves 4.37.   □

Generalization of the Fixed Point Theorem

4.39 THEOREM. Let E be a complete metric space and let T be a continuous transformation on E. If T_k is a contraction for some $k \geq 1$, then T has a unique fixed point.

PROOF. Fix k such that $U = T_k$ is a contraction. By Theorem 4.4, then, U has a unique fixed point x, and $\lim_n U_n x_0 = x$ for every point x_0 in E. Now, by the continuity of T, and since $T \circ U_n = U_n \circ T$,

$$
Tx = \lim_n TU_n x_0 = \lim_n U_n Tx_0 = x,
$$

that is, x is a fixed point of T. To show that it is the only fixed point of T we note that every fixed point of T is a fixed point of $T_k = U$, whereas U has only one fixed point, namely x.   □

Exercises

4.40   Solve the Fredholm Eq. 4.19 for arbitrary φ, on $[a, b] = [0, 2\pi]$, with the kernel

$$K(x, y) = \sin(x + y).$$

4.41   Do the same with $[a, b] = [0, 1]$ and $K(x, y) = (x - y)^2$.

4.42   Let p be a continuous function of $[0, b]$. Show that

$$f(x) = \varphi(x) + \int_0^x p(y) f(x - y)\, dy, \quad x \in [0, b],$$

has a unique solution f for each continuous function φ.

D. Differential Equations

We continue with applications of the fixed point theorem by discussing Picard's method of successive approximations for solving systems of differential equations.

We start with the simplest case where the differential equation describes the position of a particle moving on $\mathbb{R}$. The picture of the motion is given in Fig. 4.6. The motion is described by the initial data t_0 and x_0 and by a continuous function $v \colon \mathbb{R} \times \mathbb{R} \mapsto \mathbb{R}$ as follows. The particle starts from x_0 at time t_0; its velocity at time t is $v(t, x)$ if its position then is x. Thus, letting $x(t)$ denote the position of the particle at time t, we have

4.43 $$x(t) = x_0 + \int_{t_0}^{t} v(s, x(s))\, ds, \quad t \geq t_0.$$

The points t_0 and x_0 and the velocity function v are given. We are interested in the existence and uniqueness of the function x.

In the classical formulation of this problem, it is usual to express 4.43 as a differential equation:

4.44 $$\frac{dx}{dt} = v(t, x), \quad x(t_0) = x_0.$$

The following is *Picard's theorem*:

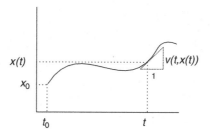

FIGURE 4.6. A moving particle.

4.45 THEOREM. Let v be defined and continuous on $[t_0, \infty) \times [a, b]$, and let x_0 be in (a, b), and suppose that v satisfies a Lipschitz condition in its spatial argument:

4.46 $$|v(t, x) - v(t, y)| \leq K|x - y|, \quad x, y \in [a, b].$$

Then, there is a number $t_1 > t_0$ such that 4.43 has a unique solution $\{x(t) : t_0 \leq t \leq t_1\}$.

PROOF. Let t_1' be an arbitrary number larger than t_0. By the continuity of v, we have

4.47 $$|v(t, x)| \leq c, \quad t_0 \leq t \leq t_1', \quad a \leq x \leq b$$

for some constant c. Choose $\delta > 0$ so that

4.48 $$K\delta < 1 \quad \text{and} \quad a \leq x_0 - c\delta < x_0 < x_0 + c\delta \leq b.$$

Let $t_1 = \min\{t_1', t_0 + \delta\}$. Let C^* be the space of all continuous functions $x \colon [t_0, t_1] \mapsto [x_0 - c\delta, x_0 + c\delta]$ with the usual supremum metric; that is, $\|x - y\| = \sup_{t_0 \leq t \leq t_1} |x(t) - y(t)|$.

The set C^* is a closed subset of the space $C([t_0, t_1] \mapsto \mathbb{R})$. Since the latter is complete, C^* is complete.

Consider the transformation T defined by letting Tx be the function

4.49 $$Tx(t) = x_0 + \int_{t_0}^{t} v(s, x(s))\, ds, \quad t \in [t_0, t_1].$$

For x in C^*, we have from 4.47 that

$$|Tx(t) - x_0| \leq \int_{t_0}^{t} |v(s, x(s))|\, ds \leq c(t - t_0) \leq c\delta,$$

which shows that $Tx \in C^*$. Moreover, for x, y in C^*,

$$
\begin{aligned}
|Tx(t) - Ty(t)| &\leq \int_{t_0}^{t} |v(s, x(s)) - v(s, y(s))|\, ds \\
&\leq \int_{t_0}^{t} K|x(s) - y(s)|\, ds \\
&\leq K\delta \|x - y\|
\end{aligned}
$$

in view of 4.46. Thus, $\|Tx - Ty\| \leq K\delta \|x - y\|$ and $K\delta < 1$ by the way δ was chosen. So, T is a contraction on C^*. Since C^* is complete, Theorem 4.4 applies to show that T has a unique fixed point x. But, $x = Tx$ means that x solves 4.43. This completes the proof.   □

The preceding can be easily generalized to the case of systems of differential equations

4.50 $$\frac{dx_i}{dt} = v_i(t, x_1, \ldots, x_n), \quad i = 1, 2, \ldots, n.$$

Before stating this generalization, we mention that the term "domain" means "an open and connected subset of a Euclidean space," and we note that 4.43 can be interpreted for $t < t_0$ by the convention that integrals from t_0 to t are the negatives of integrals from t to t_0. The following is the analog of Theorem 4.45 for motions in $\mathbb{R}^n$.

4.51 THEOREM. Let D be a domain in $\mathbb{R} \times \mathbb{R}^n$. Let v be a continuous function from D into $\mathbb{R}^n$. Suppose that $(t_0, x_0) \in D$ and that $v(t, x) = (v_1(t, x), \ldots, v_n(t, x))$ satisfies the following Lipschitz condition for some K:

4.52 $$\max_{1 \leq i \leq n} |v_i(t, x) - v_i(t, y)| \leq K \max_{1 \leq j \leq n} |x_j - y_j|.$$

Then, there is an interval $[t_0 - \delta, t_0 + \delta]$ in which the system 4.50 has a unique solution $\{x(t) : t_0 - \delta \leq t \leq t_0 + \delta\}$ satisfying $x(t_0) = x_0$.

4.53 REMARK. In integral notation, we may write 4.50 as

$$
x_i(t) = x_{0i} + \int_{t_0}^{t} v_i(s, x_1(s), \ldots, x_n(s))\, ds, \quad i = 1, \ldots, n.
$$

The claim of the preceding theorem is that this has a unique solution $\{x(t) : t_0 - \delta \leq t \leq t_0 + \delta\}$. In vector notation, we may rewrite this as

$$
x(t) = x_0 + \int_{t_0}^{t} v(s, x(s))\, ds, \quad |t - t_0| \leq \delta,
$$

which is exactly the same as 4.43 except that here $x \colon [t_0 - \delta, t_0 + \delta] \mapsto \mathbb{R}^n$ and $v \colon D \mapsto \mathbb{R}^n$.

In preparation for the proof of Theorem 4.51, let the metric on $\mathbb{R}^n$ be

$$d(x, y) = \max_{1 \le i \le n} |x_i - y_i|.$$

Then, the Lipschitz condition 4.52 can be written as

4.54 $$d(v(t, x), v(t, y)) \le K d(x, y).$$

It should be clear by now that the proof of Theorem 4.45 will go through for Theorem 4.51 as well, with some notational changes:

PROOF. By the continuity of $v_1, \ldots, v_n$, we have

$$|v_i(t, x)| \le c, \quad i = 1, \ldots, n,$$

for some $c > 0$, for all (t, x) in some domain $D' \subset D$ containing (t_0, x_0). Choose $\delta > 0$ so that

$$K\delta < 1$$

and

$$(t, x) \in D' \text{ if } t \in [t_0 - \delta, t_0 + \delta] \text{ and } d(x, x_0) \le c\delta.$$

Let C^* be the space of continuous functions $x \colon [t_0 - \delta, t_0 + \delta] \mapsto \bar{B}(x_0, c\delta)$, and let the metric on C^* be defined by

$$\|x - y\| = \max_t d(x(t), y(t)).$$

It is clear that C^* is complete. Define, for $x \in C^*$,

$$Tx(t) = x_0 + \int_{t_0}^t v(s, x(s)) \, ds, \quad t_0 - \delta \le t \le t_0 + \delta.$$

We proceed to show that T is a contraction on C^*, which will complete the proof via Theorem 4.4.

First, we show that $Tx \in C^*$ for x in C^*. For such x, it is clear that Tx is a continuous function, and

$$d(Tx(t), x_0) = \max_i \left| \int_{t_0}^t v_i(s, x(s)) \, ds \right| \le c\delta$$

for t in $[t_0 - \delta, t_0 + \delta]$ in view of the bounding of v_i by c. Thus, $Tx \in C^*$ if $x \in C^*$. Moreover, for x, y in C^*,

$$
\begin{aligned}
\|Tx - Ty\| &= \max_t d(Tx(t), Ty(t)) \\
&= \max_t \max_i \left| \int_{t_0}^t [v_i(s, x(s)) - v_i(s, y(s))]\, ds \right| \\
&\leq \max_t \int_{t_0}^t d(v(s, x(s)) - v(s, y(s)))\, ds \\
&\leq \max_t \int_{t_0}^t K d(x(s), y(s))\, ds \\
&\leq K\delta \|x - y\|,
\end{aligned}
$$

which follows from the Lipschitz condition 4.52 on v; see 4.54 as well. Since $K\delta < 1$, this shows that T is a contraction on C^*. $\square$

The preceding theorem ensures the existence and uniqueness of a solution x to the system 4.50 of differential equations. Successive approximations to x can be obtained as follows. Define

$$
\begin{aligned}
x^{(0)}(t) &= x_0, \quad t \in [t_0 - \delta, t_0 + \delta], \\
x^{(n+1)}(t) &= Tx^{(n)}(t) \\
&= x_0 + \int_{t_0}^t v(s, x^{(n)}(s))\, ds, \quad t \in [t_0 - \delta, t_0 + \delta].
\end{aligned}
$$

Then, the sequence of functions $x^{(n)}$ converges to the solution x.

Exercises

4.55 Solve the system

$$
\frac{dx_i(t)}{dt} = \sum_{j=1}^n a_{ij} x_j(t) + b_i(t), \quad i = 1, 2, \ldots, n,
$$

for smooth b and initial condition $x(0) = x_0$. How does the method of successive approximations work?

CHAPTER 5

Convexity

Linear functions are the simplest functions in linear spaces. Their study is of fundamental importance and forms the subject known as *linear algebra*. In this chapter we assume the reader has mastered the essentials of linear algebra. We therefore spend some time on the class of functions and sets that are just one step more complicated than the linear ones, namely convex functions and convex sets.

A. Convex Sets and Convex Functions

For the needs of this chapter, it will be convenient to introduce the notation

$$\mathbb{R}^* = \mathbb{R} \cup \{+\infty\}.$$

5.1 DEFINITION. A function $f \colon \mathbb{R}^n \mapsto \mathbb{R}^*$ is said to be *convex* if

5.2 $$pf(x) + qf(y) \geq f(px + qy)$$

for all x, y in $\mathbb{R}^n$ and p, q in $(0, 1)$ with $p + q = 1$. A function f on $\mathbb{R}^n$ is said to be *concave* if $-f$ is convex.

5.3 DEFINITION. A subset C of $\mathbb{R}^n$ is said to be *convex* if

$$px + qy \in C$$

for all x, y in C and $0 < p < 1$ and $q = 1 - p$.

Figure 5.1 shows a convex subset of $\mathbb{R}^2$ and a convex function on $\mathbb{R}$. Figure 5.2 gives examples of nonconvex functions and sets; the aim is to illustrate the geometric meanings of convexity.

Epigraphs

For $f \colon \mathbb{R}^n \mapsto \mathbb{R}^*$, the *epigraph* of f is the set of all pairs (x, r) in $\mathbb{R}^n \times \mathbb{R}$ satisfying $f(x) \leq r$. The epigraph relates the convexity of functions to the convexity of sets, as the next theorem shows.

E. Çınlar and R.J. Vanderbei, *Real and Convex Analysis*, Undergraduate Texts in Mathematics, DOI 10.1007/978-1-4614-5257-7_5, © Springer Science+Business Media New York 2013

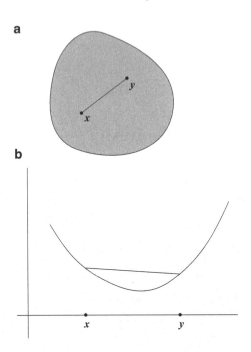

FIGURE 5.1. (**a**) A convex set. (**b**) A convex function.

5.4 THEOREM. A function $f\colon \mathbb{R}^n \mapsto \mathbb{R}^*$ is convex if and only if its epigraph is a convex subset of $\mathbb{R}^n \times \mathbb{R}$.

PROOF. Let f be convex. Fix (x, r) and (y, s) to be points in its epigraph. Fix $0 < p < 1$ and put $q = 1 - p$. Then, by 5.2,

$$f(px + qy) \le pf(x) + qf(y) \le pr + qs,$$

which means that the pair $(px + qy, pr + qs) = p(x, r) + q(y, s)$ belongs to the epigraph. Thus, the epigraph is convex.

For the converse, suppose that the epigraph of f is convex. Fix x, y in $\mathbb{R}^n$, $0 < p < 1$, $q = 1 - p$. If $f(x)$ or $f(y)$ is equal to ∞, then 5.2 holds trivially. If both $f(x)$ and $f(y)$ are real numbers, then $(x, f(x))$ and $(y, f(y))$ both belong to the epigraph of f, and thus

$$p(x, f(x)) + q(y, f(y)) = (px + qy, pf(x) + qf(y))$$

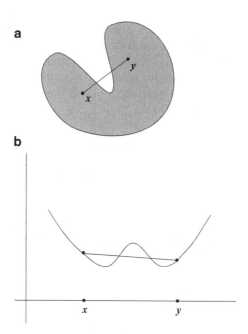

FIGURE 5.2. (**a**) A nonconvex set. (**b**) A nonconvex function.

belongs to the epigraph by the assumed convexity of the epigraph. It follows from the definition of epigraph that 5.2 holds, that is, the function f is convex.   □

Exercises

5.5 *Intersections of convex sets.* Show that the intersection of an arbitrary collection of convex sets is convex.

5.6 *Concavity.* Let $f : \mathbb{R}^n \mapsto \mathbb{R} \cup \{-\infty\}$. Show that it is concave if and only if

$$pf(x) + qf(y) \leq f(px + qy)$$

for all x, y in $\mathbb{R}^n$ and $0 < p < 1$ and $q = 1 - p$.

5.7 *Epigraph.* Show that the epigraph of f is the empty set if and only if $f = \infty$ identically.

5.8 *Supremum of convex functions.* Let f and g be convex functions on $\mathbb{R}^n$. Show that, then, $f \vee g$ is again convex. Show that the epigraph of $f \vee g$ is the intersection of the epigraphs of f and g. These properties extend to arbitrary collections (countable or uncountable) of convex functions: if $\{f_i : i \in I\}$ is a collection of convex functions on $\mathbb{R}^n$, then $f = \sup_{i \in I} f_i$ is a convex function on $\mathbb{R}^n$, and the epigraph of f is the intersection of the epigraphs of f_i, $i \in I$.

B. Projections

Throughout this section, C will remain a fixed nonempty closed convex subset of $\mathbb{R}^n$, all points are in $\mathbb{R}^n$, and $\|x - y\|$ is the usual Euclidean distance from point x to point y.

Figure 5.3 suggests that, for every point x, there is a unique point $\bar{x}$ that is the point of C closest to x. If C is a straight line in $\mathbb{R}^2$, then $\bar{x}$ would be the perpendicular projection of x onto the line C.

5.9 DEFINITION. A point $\bar{x}$ is called the *projection* of the point x onto C if $\bar{x} \in C$ and

5.10 $$\|x - \bar{x}\| = \inf_{z \in C} \|x - z\|.$$

5.11 THEOREM. Every x in $\mathbb{R}^n$ has a unique projection $\bar{x}$ on C.

PROOF. *Existence.* Fix x. Let $y \in C$, and let B be the closed ball of radius $\|x - y\|$ and center x. Note that

$$\inf_{z \in C} \|x - z\| = \inf_{z \in B \cap C} \|x - z\| = \inf_{z \in B \cap C} f(z),$$

where $f(z) = \|x - z\|$. This function f is continuous. The set $B \cap C$ is bounded since B is bounded, and is closed since both B and C are closed. Thus, $B \cap C$ is compact, and Corollary 3.24 implies that the function f must attain its infimum over that compact set. Hence, there exists $\bar{x}$ in $B \cap C \subset C$ such that 5.10 holds.

Uniqueness. Suppose that $\bar{x}$ and $\tilde{x}$ are in C and satisfy 5.10, that is

5.12 $$\|x - \bar{x}\| = \|x - \tilde{x}\| = \inf_{z \in C} \|x - z\|.$$

Put $y = \frac{1}{2}\bar{x} + \frac{1}{2}\tilde{x}$. Since C is convex, $y \in C$. Note that $x - y$ is orthogonal to $y - \bar{x}$:

$$\begin{aligned}(x - y) \cdot (y - \bar{x}) &= \left(\frac{1}{2}(x - \bar{x}) + \frac{1}{2}(x - \tilde{x})\right) \cdot \left(\frac{1}{2}(x - \bar{x}) - \frac{1}{2}(x - \tilde{x})\right) \\ &= \frac{1}{4}\|x - \bar{x}\|^2 - \frac{1}{4}\|x - \tilde{x}\|^2 = 0\end{aligned}$$

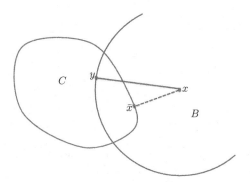

FIGURE 5.3. The projection of x on C is the point $\bar{x}$ of C.

in view of 5.12. This orthogonality implies that

$$
\begin{aligned}
\|x - \bar{x}\|^2 &= (x - y + y - \bar{x}) \cdot (x - y + y - \bar{x}) \\
&= \|x - y\|^2 + \|y - \bar{x}\|^2 \geq \|x - y\|^2.
\end{aligned}
$$

Since $\bar{x}$ satisfies 5.12, and since $y \in C$, the last inequality must be an equality, which means that $\|y - \bar{x}\| = 0$. Thus, $y = \bar{x}$, which implies that $\bar{x} = \tilde{x}$ as needed to show uniqueness. $\square$

Characterization of Projection

The following is a useful characterization of the projection $\bar{x}$ of a point x. Its geometric meaning is explained in Fig. 5.4.

5.13 THEOREM. A point $\bar{x}$ is the projection on C of the point x if and only if $\bar{x} \in C$ and

5.14 $$(x - \bar{x}) \cdot (\bar{x} - z) \geq 0, \qquad z \in C.$$

PROOF. Suppose that $\bar{x} \in C$ and 5.14 holds. For z in C, then,

$$
\begin{aligned}
\|x - z\|^2 &= (x - \bar{x} + \bar{x} - z) \cdot (x - \bar{x} + \bar{x} - z) \\
&= \|x - \bar{x}\|^2 + \|\bar{x} - z\|^2 + 2(x - \bar{x}) \cdot (\bar{x} - z).
\end{aligned}
$$

On the right side, all three terms are positive, the last one by 5.14. It follows that $\|x - \bar{x}\| \leq \|x - z\|$ for all z in C; thus, $\bar{x}$ is the projection of x on C by definition.

Suppose that $\bar{x}$ is the projection on C of the point x. Then, $\bar{x} \in C$ by definition, and there remains to show 5.14. To that end, fix z in C, and with $0 < p < 1$ put

$$y = pz + (1 - p)\bar{x}.$$

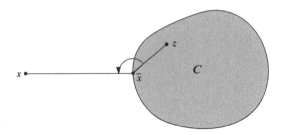

FIGURE 5.4. The vector from $\bar{x}$ to x makes an obtuse angle with the vector from $\bar{x}$ to every other point z of C.

Then, $y \in C$ by the convexity of C, and by the definition of $\bar{x}$,

$$\|x - \bar{x}\|^2 \leq \|x - y\|^2 = \|x - \bar{x} + p(\bar{x} - z)\|^2$$
$$= \|x - \bar{x}\|^2 + 2p(x - \bar{x}) \cdot (\bar{x} - z) + p^2 \|\bar{x} - z\|^2.$$

Hence,

$$2(x - \bar{x}) \cdot (\bar{x} - z) + p\|\bar{x} - z\|^2 \geq 0.$$

Since $p > 0$ can be taken arbitrarily small, this implies 5.14. $\square$

5.15 COROLLARY. Let $\bar{x} \in C$ and suppose that $z \cdot x = z \cdot \bar{x}$ for every z in C. Then, $\bar{x}$ is the projection of x on C.

PROOF. Under the assumption, $\bar{x} \cdot x = \bar{x} \cdot \bar{x}$ as well (take $z = \bar{x}$). Thus, for every z in C,

$$(x - \bar{x}) \cdot (\bar{x} - z) = x \cdot \bar{x} - \bar{x} \cdot \bar{x} - x \cdot z + \bar{x} \cdot z = 0,$$

and the claim follows from the preceding theorem. $\square$

The following gives a converse to the preceding corollary in the special case where C is a linear subspace of $\mathbb{R}^n$. In addition, it lists an explicit formula for computing the projection.

5.16 THEOREM. Let A be an $m \times n$ matrix of rank m. Suppose that

$$C = \{z \in \mathbb{R}^n : z = A^T y \text{ for some } y \text{ in } \mathbb{R}^m\}.$$

Then, the following are equivalent:
  (a) $\bar{x}$ is the projection of x onto C.
  (b) $\bar{x} = A^T(AA^T)^{-1}Ax$.
  (c) $\bar{x} \in C$ and $x \cdot z = \bar{x} \cdot z$ for every z in C.

REMARK. The matrix A has m rows; each row is a vector in $\mathbb{R}^n$; the rows are linearly independent since A has rank m. The assumption is that C is the linear space spanned by these vectors, that is, each point of C is a linear combination of the rows of A. Also, since A has rank m, the $m \times m$ matrix AA^T is nonsingular, that is, the inverse $(AA^T)^{-1}$ exists.

PROOF. (a) $\Rightarrow$ (b): Suppose (a). By the definition of projections, $\bar{x} \in C$ and $\|x - \bar{x}\|^2 \leq f(y)$ for every y in $\mathbb{R}^m$, where

$$f(y) = \|x - A^T y\|^2 = x \cdot x - 2x \cdot (A^T y) + (A^T y) \cdot (A^T y).$$

Let $\bar{y}$ be a point in $\mathbb{R}^m$ where the gradient of f vanishes, that is,

$$\nabla f(\bar{y}) = -2Ax + 2AA^T \bar{y} = 0.$$

Since AA^T is nonsingular, $\bar{y}$ is unique and is given by

$$\bar{y} = (AA^T)^{-1} Ax.$$

Hence, $\bar{x} = A^T \bar{y}$ is as claimed by (b).

(b) $\Rightarrow$ (c): Let $\bar{x}$ be as given in (b). Then, $\bar{x} = A^T \bar{y}$, with an obvious definition for $\bar{y}$ in $\mathbb{R}^m$; hence $\bar{x} \in C$. If $z \in C$, then $z = A^T y$ for some y in $\mathbb{R}^m$, and

$$z \cdot \bar{x} = (y^T A) A^T (AA^T)^{-1} Ax = (y^T A)x = z \cdot x$$

as claimed by (c).

(c) $\Rightarrow$ (a): by Corollary 5.15. $\square$

C. Supporting Hyperplane Theorem

A *half-space* H is a set of the form

5.17 $$H = \{z \in \mathbb{R}^n : \xi \cdot z \leq b\},$$

where b is a fixed real number and ξ is a fixed nonzero vector in $\mathbb{R}^n$. Then, its *boundary* ∂H is the hyperplane

5.18 $$\partial H = \{z \in \mathbb{R}^n : \xi \cdot z = b\}.$$

5.19 THEOREM. Let C be a nonempty closed convex set in $\mathbb{R}^n$. Let x be a point in $\mathbb{R}^n \setminus C$. Then, there exists a half-space H such that

$$C \subset H, \qquad C \cap \partial H \neq \emptyset, \qquad x \notin H.$$

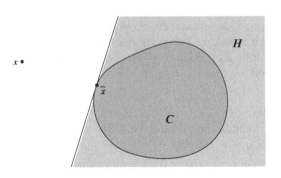

FIGURE 5.5. The supporting hyperplane theorem.

PROOF. Let $\bar{x}$ denote the projection of x on C. Put $\xi = x - \bar{x}$. Since $\bar{x}$ is in C and x is outside, $\xi \neq 0$. Thus,

$$H = \{z \in \mathbb{R}^n : \xi \cdot z \leq \xi \cdot \bar{x}\}$$

is a half-space. If $z \in C$, then Theorem 5.13 shows that $z \in H$; thus, $C \subset H$. Since $\xi \cdot x - \xi \cdot \bar{x} = \|\xi\|^2 > 0$, it follows that x is outside H. Finally, since $\bar{x} \in C$ and $\bar{x} \in \partial H$, the intersection $C \cap \partial H$ is not empty (Fig. 5.5). $\qquad \square$

Exercises

5.20 *Supporting hyperplanes.* Let $f : \mathbb{R}^n \mapsto \mathbb{R}^*$ be a convex function. Show that f is the supremum of all affine functions it dominates. That is,

$$f(x) = \sup_{h \in \mathcal{H} : h \leq f} h(x),$$

where

$$\mathcal{H} = \{h : h(x) = \xi \cdot x - b \text{ for some } \xi \in \mathbb{R}^n, b \in \mathbb{R}\}.$$

D. Legendre Transform

Let $f : \mathbb{R}^n \mapsto \mathbb{R}^*$ and let $\xi \in \mathbb{R}^n$ be fixed. Suppose that f dominates the affine function $x \mapsto \xi \cdot x - b$, that is, $f(x) \geq \xi \cdot x - b$ for all x in $\mathbb{R}^n$. Then,

$$\xi \cdot x - f(x) \leq b, \qquad x \in \mathbb{R}^n,$$

that is, b is an upper bound for the function $x \mapsto \xi \cdot x - f(x)$. When f is convex, the smallest such bound is of fundamental importance, especially as a function of ξ.

5.21 DEFINITION. Let $f: \mathbb{R}^n \mapsto \mathbb{R}^*$ be convex. Its *Legendre transform* is the function $\widehat{f}: \mathbb{R}^n \mapsto \mathbb{R}^*$ defined by

$$\widehat{f}(\xi) = \sup_{x \in \mathbb{R}^n} (\xi \cdot x - f(x)), \qquad \xi \in \mathbb{R}^n.$$

Examples

We shall see shortly that the Legendre transform $\widehat{f}$ of a convex function f is again convex. Here are some explicit examples.

5.22 Let $f: \mathbb{R} \mapsto \mathbb{R}$ be given by $f(x) = |x|$. Then, for ξ in $\mathbb{R}$,

$$\widehat{f}(\xi) = \begin{cases} 0 & \text{if } |\xi| \leq 1, \\ \infty & \text{otherwise.} \end{cases}$$

5.23 Let $f(x) = ax$, $x \in \mathbb{R}$, where $a \in \mathbb{R}$ is fixed. Then, for ξ in $\mathbb{R}$,

$$\widehat{f}(\xi) = \begin{cases} 0 & \text{if } \xi = a, \\ \infty & \text{otherwise.} \end{cases}$$

5.24 Let $f(x) = ax^2$, $x \in \mathbb{R}$, with $a > 0$ fixed. Then, for $\xi \in \mathbb{R}$,

$$\widehat{f}(\xi) = \frac{1}{4a}\xi^2.$$

5.25 Let $a \in \mathbb{R}^n$, $b \in \mathbb{R}$, both fixed. Let $f(x) = a \cdot x - b$, $x \in \mathbb{R}^n$. Then,

$$\widehat{f}(\xi) = \begin{cases} b & \text{if } \xi = a, \\ \infty & \text{otherwise.} \end{cases}$$

5.26 Let a and b be as in the preceding example, and, for x in $\mathbb{R}^n$, let $f(x) = b$ if $x = a$ and $f(x) = \infty$ otherwise. Then,

$$\widehat{f}(\xi) = a\xi - b, \qquad \xi \in \mathbb{R}^n.$$

Involution

As the examples above illustrate, the Legendre transform $\widehat{f}$ of a convex function f is again convex, and the transform of $\widehat{f}$ is the original f. Before stating this as a theorem, we set forth next an elementary fact or two.

5.27 LEMMA. Let f and g be convex functions on $\mathbb{R}^n$.

  (a) If $f \le g$, then $\widehat{f} \ge \widehat{g}$.
  (b) If $f = cg$ for some constant $c > 0$, then $\widehat{f}(\xi) = c\widehat{g}(\xi/c)$.

PROOF. If $f \le g$, then

$$\widehat{f}(\xi) = \sup_x (\xi \cdot x - f(x)) \ge \sup_x (\xi \cdot x - g(x)) = \widehat{g}(\xi).$$

If $f = cg$ with $c > 0$ constant, then

$$\widehat{f}(\xi) = \sup_x (\xi \cdot x - cg(x))$$

$$= c\sup_x \left(\left(\frac{1}{c}\xi \right) \cdot x - g(x) \right) = cg\left(\frac{1}{c}\xi \right).$$

□

  The following is the promised theorem. Recall that a function is convex if and only if its epigraph is convex.

5.28 THEOREM. **(Involution)** If the epigraph of f is nonempty, closed, and convex, then so is the epigraph of its Legendre transform $\widehat{f}$, and, moreover, $\widehat{\widehat{f}} = f$.

PROOF. Let the epigraph of f be as described. The supporting hyperplane theorem implies that there exists an affine function $h \le f$, say $h(x) = a \cdot x - b$. Also, since the epigraph is not empty, there is a point x^* such that $f(x^*) = c < \infty$; define $g(x)$ to be equal to c if $x = x^*$, and to ∞ otherwise. Then, $h \le f \le g$, which implies via Proposition 5.27 that

5.29 $\widehat{g} \le \widehat{f} \le \widehat{h},$

whereas, as Examples 5.25 and 5.26 show,

$$\widehat{g}(\xi) = \xi \cdot x^* - c, \qquad \widehat{h}(\xi) = \begin{cases} b & \text{if } \xi = a, \\ \infty & \text{otherwise.} \end{cases}$$

The first inequality in 5.29 implies that $\widehat{f}$ cannot take $-\infty$ as a value; that is, $\widehat{f} : \mathbb{R}^n \mapsto \mathbb{R}^*$. The second inequality implies that $\widehat{f}$ takes some real values. Hence, the epigraph of $\widehat{f}$ is nonempty.
  Next, observe that

$$\widehat{f}(\xi) = \sup_x (\xi \cdot x - f(x)) = \sup\{\xi \cdot x - b : b \ge f(x), x \in \mathbb{R}^n\},$$

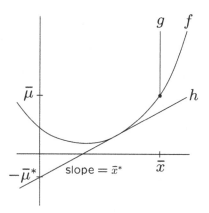

FIGURE 5.6. The functions f, g, and h.

that is, the epigraph of $\widehat{f}$ is the intersection of the epigraphs of all the affine functions $x \mapsto \xi \cdot x - b$ dominated by f. The epigraph of an affine function is a closed half-space, which is closed and convex. The intersection of an arbitrary family of closed sets is closed (Theorem 2.34), and similarly, an arbitrary intersection of convex sets is convex (Exercise 5.5). Hence, the epigraph of $\widehat{f}$ is closed and convex (Fig. 5.6).

Finally, by Exercise 5.20,

$$f(x) = \sup\{\xi \cdot x - b : \xi \in \mathbb{R}^n, b \ge \widehat{f}(\xi)\} = \sup_{\xi} \left(\xi \cdot x - \widehat{f}(\xi) \right) = \widehat{\widehat{f}}(x).$$

$\square$

Fenchel's Inequality

5.30 PROPOSITION. Let f be convex. Then $\xi \cdot x \le f(x) + \widehat{f}(\xi)$, for every ξ and x.

PROOF. By the definition of the Legendre transform, for every ξ, $\widehat{f}(\xi) \ge \xi \cdot x - f(x)$, which is equivalent to the claim. $\square$

Subgradient

5.31 DEFINITION. Let $f \colon \mathbb{R}^n \mapsto \mathbb{R}^*$ and $\xi \in \mathbb{R}^n$. Then, ξ is a *subgradient* of f at x if

$$f(z) \geq f(x) + \xi \cdot (z - x)$$

for z in $\mathbb{R}^n$, that is, if $z \mapsto f(x) + \xi \cdot (z - x)$ is a supporting hyperplane for the epigraph of f. Define

$$\partial f(x) = \{\xi \in \mathbb{R}^n : \xi \text{ is a subgradient of } f \text{ at } x\}.$$

It is easy to check that $\partial f(x)$ is a closed convex subset of $\mathbb{R}^n$. The following gives some geometric meanings and connections to the Legendre transform.

5.32 THEOREM. Let $f \colon \mathbb{R}^n \mapsto \mathbb{R}^*$ be convex and fix x in $\mathbb{R}^n$. Then the following are equivalent:

(a) $\xi \in \partial f(x)$.
(b) $\xi \cdot z - f(z)$ attains its maximum at $z = x$.
(c) $f(x) + \widehat{f}(\xi) = \xi \cdot x$.
(d) $x \in \partial \widehat{f}(\xi)$.
(e) $\lambda \cdot x - \widehat{f}(\lambda)$ attains its maximum at $\lambda = \xi$.

PROOF. By definition, ξ is a subgradient if and only if

$$\xi \cdot x - f(x) \geq \xi \cdot z - f(z),$$

for all $z \in \mathbb{R}^n$. Thus, since $\widehat{f}(\xi)$ is the supremum of the right side,

$$\xi \cdot x - f(x) = \widehat{f}(\xi).$$

Hence, (a), (b), and (c) are equivalent. By the symmetry of (c), then, (c), (d), and (e) are equivalent.   □

Examples

Returning to the examples from $\mathbb{R}^1$, we make the following subgradient calculations.

5.33 Let f and $\widehat{f}$ be as in Example 5.23. Then, it follows from Theorem 5.32(c) that $\partial f(x) = \{a\}$, which agrees with the usual definition of the derivative of f.

5.34 Let f and $\widehat{f}$ be as in Example 5.22. Now, the equality given by Theorem 5.32(c) can be satisfied if and only if $x\xi = |x|$ and $|\xi| \le 1$. These two conditions are equivalent to

$$\xi = \begin{cases} 1 & x > 0, \\ -1 & x < 0, \\ [-1, 1] & x = 0. \end{cases}$$

Thus, again $\partial f(x)$ agrees with the usual definition of the derivative of f, at least for $x \ne 0$. And, for $x = 0$, the function is not differentiable in the normal sense but has a subdifferential that is given by the interval $[-1, 1]$.

5.35 Let f and $\widehat{f}$ be as in Example 5.24. Now, part (c) of Theorem 5.32 reads $x\xi = f(x) + \widehat{f}(\xi) = ax^2/2 + \xi^2/(2a)$. This equality can be satisfied if and only if $\xi^2 - 2ax\xi + a^2x^2 = 0$. This quadratic form is easily factored:

$$(\xi - ax)^2 = 0.$$

Hence, $\xi = ax$, which, again, is consistent with the usual definition of derivative.

Exercises

5.36 For each of the following functions, compute its Legendre transform and check whether $\widehat{\widehat{f}} = f$.

(a) $f(x) = e^x$.

(b) Assuming that $1 < p < \infty$, $f(x) = |x|^p/p$.

(c) Assuming $a \ge 0$,

$$f(x) = \begin{cases} -(a^2 - x^2)^{1/2} & \text{if } |x| \le a, \\ \infty & \text{otherwise} \end{cases}$$

(see Fig. 5.7).

(d) Assuming that $0 < a < b$,

$$f(x) = \begin{cases} |x| & |x| < a, \\ b & |x| = a, \\ \infty & |x| > a. \end{cases}$$

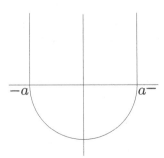

FIGURE 5.7. The functions f in Exercise 5.36 part (c).

E. Infimal Convolution

This is similar, conceptually, to the convolution of two functions. We shall see, also, that the Legendre transform plays the same role here as the Fourier transform does in that classical setting.

5.37 DEFINITION. Let f and g be functions defined on $\mathbb{R}^n$. Their *infimal convolution*, denoted $f \star g$, is defined by

$$f \star g(x) = \inf_{y \in \mathbb{R}^n} \left(f(y) + g(x - y) \right).$$

It is checked easily that the operation is commutative and associative (see Exercise 5.40) and thus extends to any finite number of functions: the infimal convolution f of $f_1, \ldots, f_m$ is given by

$$f(x) = \inf \left\{ f_1(x_1) + \cdots + f_m(x_m) : x_1, \ldots, x_m \in \mathbb{R}^n, x_1 + \cdots + x_m = x \right\}.$$

5.38 THEOREM. The Legendre transform of an infimal convolution is the sum of the Legendre transforms, that is,

$$\widehat{f_1 \star \cdots \star f_m} = \widehat{f_1} + \cdots + \widehat{f_m}.$$

PROOF. It is sufficient to prove this for $m = 2$. Let $f = g \star h$. Fix $\xi \in \mathbb{R}^n$. Using the formula for the infimal convolution, we have

$$
\begin{aligned}
\widehat{f}(\xi) &= \sup_{x} \left(\xi \cdot x - \inf_{y,z:y+z=x} (g(y) + h(z)) \right) \\
&= \sup_{x} \sup_{y,z:y+z=x} (\xi \cdot x - g(y) - h(z)) \\
&= \sup_{y,z} (\xi \cdot y - g(y) + \xi \cdot z - h(z)) \\
&= \sup_{y} (\xi \cdot y - g(y)) + \sup_{z} (\xi \cdot z - h(z)) = \widehat{g}(\xi) + \widehat{h}(\xi).
\end{aligned}
$$

$\square$

5.39 COROLLARY. If the epigraphs of $f_1, \ldots, f_m$ are nonempty, closed, and convex, and the epigraph of $\widehat{f}_1 \star \cdots \star \widehat{f}_m$ also satisfies these same three conditions, then

$$
\overline{f_1 + \cdots + f_m} = \widehat{f}_1 \star \cdots \star \widehat{f}_m.
$$

PROOF. Again, it suffices to prove the result for $m = 2$. If the epigraphs of g and h are nonempty, closed, and convex, then the involution theorem implies that

$$
\widehat{\widehat{g} \star \widehat{h}} = \widehat{\widehat{g} + \widehat{h}} = g + h.
$$

If the epigraph of $\widehat{g} \star \widehat{h}$ is nonempty, closed, and convex, then

$$
\overline{g + h} = \widehat{g} \star \widehat{h}.
$$

$\square$

Exercises

5.40 Show that $f \star g = g \star f$ and that $(f \star g) \star h = f \star (g \star h)$.

CHAPTER 6

Convex Optimization

This chapter is about the problem of optimizing a convex function. The concept of duality is introduced and connections with Lagrangians are made. Throughout, a convex function is said to be proper if its epigraph is not empty, that is, the function does take some real values. Also, by an abuse of language, it will be said to be *closed* if its epigraph is closed.

A. Primal and Dual Problems

The basic problem of convex optimization is that of finding the infimum of a convex function on $\mathbb{R}^n$. It turns out to be highly beneficial to enlarge the problem somewhat and to consider the problem of finding the infimum of $f(x,0)$ over all x, where f is a convex function on $\mathbb{R}^n \times \mathbb{R}^m$. Because the last m variables are set to zero, this is still the problem of minimizing a convex function over $\mathbb{R}^n$. But, the presence of the last m variables enables us to perturb the original problem in a systematic fashion. In turn, this leads to a maximization problem in the dual space $\mathbb{R}^m$, and it is best to consider the two problems together.

6.1 DEFINITION. Let f be a proper, closed, convex function on $\mathbb{R}^n \times \mathbb{R}^m$ and let

6.2 $\quad f^*(\xi, \lambda) = \inf_{x \in \mathbb{R}^n, y \in \mathbb{R}^m} \left(-\xi \cdot x - \lambda \cdot y + f(x,y) \right), \qquad \xi \in \mathbb{R}^n, \ \lambda \in \mathbb{R}^m.$

The *primal problem* is that of finding

$$\inf_{x \in \mathbb{R}^n} f(x,0),$$

and the *dual problem* is that of finding

$$\sup_{\lambda \in \mathbb{R}^m} f^*(0, \lambda).$$

Throughout this section, f and f^* will be as described in the preceding definition. Because of their roles, they are called the *primal* and *dual functions*.

Note that the dual function f^* is the negative of the Legendre transform of f, that is, $f^* = -\widehat{f}$ in the notation of Definition 5.21. Thus, it follows from the involution theorem (Theorem 5.28) that $-f^*$ is proper, closed, and convex, and $f = \widehat{-f^*}$

E. Çınlar and R.J. Vanderbei, *Real and Convex Analysis*, Undergraduate Texts in Mathematics, DOI 10.1007/978-1-4614-5257-7_6, © Springer Science+Business Media New York 2013

(the Legendre transform of $-f^*$). In short, *dual of dual equals primal*. We state this next.

6.3 THEOREM. The dual function f^* is proper, closed, and concave and

$$f(x,y) = \sup_{\xi,\lambda} \left(\xi \cdot x + \lambda \cdot y + f^*(\xi,\lambda) \right), \qquad x \in \mathbb{R}^n, \ y \in \mathbb{R}^m.$$

Weak Duality

Interest in duality stems from the so-called *duality theorems*. There are two of them: weak and strong. We start with the weak duality theorem; it is simple and easy to prove.

6.4 THEOREM. We have $f^*(0,\lambda) \leq f(x,0)$ for every λ in $\mathbb{R}^m$ and x in $\mathbb{R}^n$.

PROOF. In the definition 6.2, put $\xi = 0$ and replace x with w; we get

$$f^*(0,\lambda) = \inf_{w,y} \left(-\lambda \cdot y + f(w,y) \right) \leq \inf_{w} f(w,0) \leq f(x,0)$$

for arbitrary λ and x. $\qquad\qquad\qquad\qquad\qquad\qquad\qquad\qquad\qquad\qquad$ □

The preceding proof treated the primal and dual functions in an asymmetric manner. A better proof is via Fenchel's inequality (see Proposition 5.30 and recall that $f^* = -\widehat{f}$):

$$\xi \cdot x + \lambda \cdot y \leq f(x,y) - f^*(\xi,\lambda).$$

Now, put $\xi = 0$ and $y = 0$ to obtain $f^*(0,\lambda) \leq f(x,0)$ once again.

An equivalent way of stating the weak duality theorem is that

6.5 $$\sup_{\lambda} f^*(0,\lambda) \leq \inf_{x} f(x,0).$$

The difference between the two sides is called the *duality gap*.

Strong Duality

The strong duality theorem asserts, under some mild conditions, that the *duality gap vanishes*, that is,

6.6 $$\sup_{\lambda} f^*(0,\lambda) = \inf_{x} f(x,0),$$

or, in other words, the value of the optimal solution to the primal problem is the same as that to the dual problem.

We shall lead to the theorem via several lemmas; our aim is to reveal the reasons for the needed conditions. We start by introducing two functions which play important roles, $\varphi \colon \mathbb{R}^m \mapsto \bar{\mathbb{R}}$ and $\varphi^* \colon \mathbb{R}^n \mapsto \bar{\mathbb{R}}$ (recall that $\bar{\mathbb{R}} = \mathbb{R} \cup \{-\infty, +\infty\}$), by letting

6.7
$$\varphi(y) = \inf_{x \in \mathbb{R}^n} f(x, y), \qquad \varphi^*(\xi) = \sup_{\lambda \in \mathbb{R}^m} f^*(\xi, \lambda).$$

The development below uses φ; the arguments with φ^* are similar.

6.8 LEMMA. If φ is proper, closed, and convex, then the duality gap vanishes.

PROOF. Suppose that φ is such. By the definition 6.2 of the dual function f^*,

$$f^*(0, \lambda) = \inf_{x,y} (-\lambda \cdot y + f(x, y)) = \inf_{y} (-\lambda \cdot y + \varphi(y)).$$

Since φ is convex, the last expression here is the definition of $-\widehat{\varphi}$, where $\widehat{\varphi}$ is the Legendre transform of φ. Since φ is proper, closed, and convex, the involution theorem (Theorem 5.28) applies: φ is the Legendre transform of $\widehat{\varphi} = -f^*(0, \lambda)$, that is,

$$\varphi(y) = \sup_{\lambda} (\lambda \cdot y - \widehat{\varphi}(\lambda)) = \sup_{\lambda} (\lambda \cdot y + f^*(\lambda)).$$

Putting $y = 0$ here and in 6.7 shows that 6.6 holds, that is, the duality gap vanishes.
□

The preceding lemma reduces strong duality to the condition that φ be proper, closed, and convex. The following reduces the condition further.

6.9 LEMMA. The function φ is proper. It is also convex provided that $\varphi(y) > -\infty$ for all y in $\mathbb{R}^m$.

PROOF. Since f is proper, there is a point (x', y') such that $f(x', y') < \infty$, which implies by 6.7 that $\varphi(y') < \infty$; thus, φ is proper. Suppose, next, that $\varphi(y) > -\infty$ for all y; then, φ maps $\mathbb{R}^m$ into $\mathbb{R}^* = \mathbb{R} \cup \{+\infty\}$ as necessary for φ to be convex. As to the convex inequality, fix (x, y) and (w, z) in $\mathbb{R}^n \times \mathbb{R}^m$ and let $0 < p, q < 1$ with $p + q = 1$. Then,

$$\begin{aligned} \varphi(py + qz) &\leq f(px + qw, py + qz) \\ &\leq f(p(x, y) + q(w, z)) \leq pf(x, y) + qf(w, z), \end{aligned}$$

where the last inequality used the convexity of f. Now, holding y and z fixed, take the infimum of the rightmost expression over x and w; the result is the convexity inequality for φ. □

The next definition paves the way to checking whether $\varphi(y) > -\infty$ for all y, and thus to convexity of φ.

6.10 DEFINITION. The primal function f is said to be *feasible* if there exists some x such that $f(x, 0) < \infty$. The dual function f^* is said to be *feasible* if there exists some λ such that $f^*(0, \lambda) > -\infty$.

6.11 LEMMA. If f^* is feasible, then φ is proper and convex. If f is feasible, then φ^* is proper and concave.

PROOF. We prove the first statement only; the second is similar. Suppose that f^* is feasible, and pick $\bar{\lambda}$ such that $f^*(0, \bar{\lambda}) > -\infty$. By Theorem 6.3,

$$f(x, y) = \sup_{\xi, \lambda} (\xi \cdot x + \lambda \cdot y + f^*(\xi, \lambda)) \geq \bar{\lambda} \cdot y + f^*(0, \bar{\lambda})$$

for all x and y. Hence,

$$\varphi(y) = \inf_x f(x, y) \geq \bar{\lambda} \cdot y + f^*(0, \bar{\lambda}) \geq -\infty$$

for all y. It follows from Lemma 6.9 that φ is proper and convex. □

Combining the lemmas above, we obtain the following version of the strong duality.

6.12 THEOREM. (a) If f^* is feasible and φ is closed, then the primal and dual problems have the same optimal value, that is,

$$\inf_x f(x, 0) = \sup_\lambda f^*(0, \lambda);$$

the common value may be $+\infty$.

(b) If f is feasible and φ^* is closed, then the same equality holds; the common value may be $-\infty$.

The subtle point is that φ or φ^* might fail to be closed; see Exercises 6.30 and 6.34 for possible examples. The duality theorem above has a strong conclusion, but its utility is marred by the difficulty of checking whether φ or φ^* is closed. We seek simpler conditions next.

Closedness

Recall that the primal function f is proper, closed, and convex, which is the same as saying that its epigraph is a nonempty, closed, convex subset of $\mathbb{R}^n \times \mathbb{R}^m \times \mathbb{R}$. We are interested in checking whether the function φ is closed, that is, whether its epigraph is a closed subset of $\mathbb{R}^m \times \mathbb{R}$.

The next theorem relates the two epigraphs, epi f and epi φ for short.

6.13 THEOREM. Let D be the natural projection of epi f into $\mathbb{R}^m \times \mathbb{R}$, that is, let

$$D = \{(y, r) \in \mathbb{R}^m \times \mathbb{R} : (x, y, r) \in \text{epi } f \text{ for some } x \text{ in } \mathbb{R}^n\},$$

and let $\bar{D}$ be its closure. Then, $D \subset \text{epi } \varphi \subset \bar{D}$.

PROOF. Let $(y, r) \in D$. Then, $(x, y, r) \in \text{epi } f$ for some x, which means that $f(x, y) \leq r$ for some x. Thus,

6.14
$$\varphi(y) = \inf_x f(x, y) \leq r,$$

which is equivalent to saying that $(y, r) \in \text{epi } \varphi$. Hence, $D \subset \text{epi } \varphi$.

To prove the other containment, let $(y, r) \in \text{epi } \varphi$, which is equivalent to saying that 6.14 holds. If the infimum in 6.14 is attained, there is an x such that $f(x, y) = \varphi(y) \leq r$, which implies that $(x, y, r) \in \text{epi } f$, which in turn yields $(y, r) \in D \subset \bar{D}$, and the proof is done. Next, suppose that the infimum is not attained, and pick a sequence (ϵ_k) of numbers strictly decreasing to 0. By the definition of infimum, for each k there is a vector x_k such that

$$f(x_k, y) \leq \varphi(y) + \epsilon_k \leq r + \epsilon_k.$$

Then, $(x_k, y, r + \epsilon_k) \in \text{epi } f$, and thus, $(y, r + \epsilon_k) \in D$ for every k. Letting $k \to \infty$, we see that $(y, r) \in \bar{D}$. Hence, epi $\varphi \subset \bar{D}$.   □

6.15 COROLLARY. If D is closed, then φ is closed.

PROOF. If D is closed, $D = \bar{D}$, which means that the conclusion of the preceding theorem becomes epi $\varphi = \bar{D}$, which is closed. $\qquad\square$

In the next section, we take up linear programming, where the shape of f is such that D is always closed.

B. Linear Programming and Polyhedra

6.16 DEFINITION. A *polyhedron* in $\mathbb{R}^n$ is a set P of the form

$$P = \{x \in \mathbb{R}^n : Ax \leq b\},$$

where A is an $m \times n$ matrix and b is a vector in $\mathbb{R}^m$.

6.17 THEOREM. Every polyhedron is closed.

PROOF. The inequality $Ax \leq b$ is in fact a set of inequalities of the form $\xi_i \cdot x \leq b_i$, where ξ_i is the i-row of A. Since $\{x : \xi_i \cdot x \leq b_i\}$ is a half-space for each i, we see that a polyhedron is the intersection of some number m of half-spaces. Each half-space is closed by Theorem 2.49, and the intersection of closed sets is closed. So, every polyhedron is closed. $\qquad\square$

Consider a polyhedron. If it is a bounded set, then it can be characterized as the set of all convex combinations of its vertices. If it is unbounded, then there must be some edges that extend to infinity. In this case, the polyhedron cannot be characterized solely in terms of its vertices—one must also include positive combinations of its unbounded edges (see Fig. 6.1). These comments provide the intuition behind the following theorem, which we offer without proof. It is known as the *Minkowski–Weyl* theorem or *finite basis* theorem.

6.18 THEOREM. (Finite basis)
(a) Given a matrix A and a vector b, there exists a finite set of vectors v_j and a finite set of vectors w_k such that

6.19 $\quad \{x : Ax \leq b\} = \left\{\sum_j \alpha_j v_j + \sum_k \beta_k w_k : \alpha_j \geq 0, \sum_j \alpha_j = 1, \beta_k \geq 0\right\}.$

(b) Given a finite set of vectors v_j and another finite set of vectors w_k, there exist a matrix A and a column vector b such that 6.19 holds.

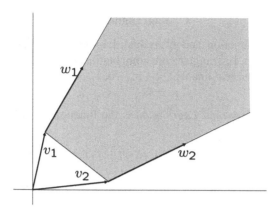

FIGURE 6.1. The gray region is a polyhedron formed by the intersection of three half planes. The vectors v_1, v_2, w_1, and w_2 from the finite basis theorem are shown.

6.20 THEOREM. Let P be a polyhedron in $\mathbb{R}^n$, and B a matrix of dimension $m \times n$. Then,

$$BP = \{Bx \: : \: x \in P\}$$

is a polyhedron in $\mathbb{R}^m$.

PROOF. By the finite basis theorem, there are v_j's and w_k's in $\mathbb{R}^n$ such that 6.19 holds. Then, the Bv_j's and the Bw_k's are vectors in $\mathbb{R}^m$, and

$$BP = \left\{ \sum_j \alpha_j Bv_j + \sum_k \beta_k Bw_k \: : \: \alpha_j \geq 0, \: \sum_j \alpha_j = 1, \: \beta_k \geq 0 \right\},$$

which is a polyhedron by the sufficiency part of 6.19.   □

We apply the theorems above on polyhedra to settle the matter of closedness. Recall the primal function f on $\mathbb{R}^n \times \mathbb{R}^m$, define φ as before by 6.7, and recall Theorem 6.13 and its Corollary 6.15. Note that the set D of Theorem 6.13 has the form BP where $P = \text{epi } f$, and B is a matrix of dimensions $m + 1$ by $m + n + 1$. Hence, as a corollary to the preceding theorem, we get the following.

6.21 THEOREM. If epi f is a polyhedron, then φ is closed.

In the case of linear programs, we have seen that epi f is a polyhedron. Thus, strong duality follows provided only that either the primal or the dual be feasible.

C. Lagrangians

Let f be a primal function, and f^* its dual. Lagrangians represent halfway points between them. As such, Lagrangians are computationally convenient. Recall the notation $\bar{\mathbb{R}}$ for the extended real line, $\mathbb{R} \cup \{+\infty, -\infty\}$.

6.22 DEFINITION. The *primal Lagrangian* is the function $L \colon \mathbb{R}^n \times \mathbb{R}^m \mapsto \bar{\mathbb{R}}$ defined by

$$L(x, \lambda) = \inf_y \left(-\lambda \cdot y + f(x, y) \right),$$

and the *dual Lagrangian* is the function $L^* \colon \mathbb{R}^n \times \mathbb{R}^m \mapsto \bar{\mathbb{R}}$ given by

$$L^*(x, \lambda) = \sup_\xi \left(\xi \cdot x + f^*(\xi, \lambda) \right).$$

Here is the point of introducing them.

6.23 THEOREM. Let L be the primal Lagrangian, and L^* the dual. Then,

$$
\begin{aligned}
f^*(\xi, \lambda) &= \inf_x \left(-\xi \cdot x + L(x, \lambda) \right), \\
f(x, y) &= \sup_\lambda \left(\lambda \cdot y + L^*(x, \lambda) \right).
\end{aligned}
$$

PROOF. Putting together the definitions of f^* and L, we get

$$
\begin{aligned}
f^*(\xi, \lambda) &= \inf_{x,y} \left(-\xi \cdot x - \lambda \cdot y + f(x, y) \right) \\
&= \inf_x \left(-\xi \cdot x + \inf_y \left(-\lambda \cdot y + f(x, y) \right) \right) = \inf_x \left(-\xi \cdot x + L(x, \lambda) \right).
\end{aligned}
$$

as claimed. The claim regarding f and L^* is proved similarly.   □

One might hope, or even expect, that L and L^* are the same function—after all, both represent halfway points between f and f^*. This is true, but subject to some conditions—see Example 6.26 for its failure.

6.24 THEOREM.
   (a) For fixed λ in $\mathbb{R}^m$, suppose that the function $x \mapsto L(x, \lambda)$ is proper, closed, and convex; then $L(x, \lambda) = L^*(x, \lambda)$ for all x in $\mathbb{R}^n$.
   (b) For each x in $\mathbb{R}^n$, suppose that the function $\lambda \mapsto L(x, \lambda)$ is proper, closed, and concave; then $L(x, \lambda) = L^*(x, \lambda)$ for all λ in $\mathbb{R}^m$.

PROOF. We prove the first statement; the second is analogous. Fix λ, let $g(x) = L(x, \lambda)$, and assume that g is proper, closed, and convex. By Theorem 6.23, we have $f^*(\,\cdot\,, \lambda) = -\widehat{g}$, the negative of the Legendre transform of g. Since g is proper, closed, and convex by assumption, Theorem 5.28 applies: g is the Legendre transform of $-f^*(\,\cdot\,, \lambda)$. But this last Legendre transform is equal to $L^*(\,\cdot\,, \lambda)$ by Definition 6.22. Hence, $L(\,\cdot\,, \lambda)$ and $L^*(\,\cdot\,, \lambda)$ are the same function. From Theorem 6.23, we have

$$f^*(\,\cdot\,, \lambda) = -\widehat{L(\,\cdot\,, \lambda)}.$$

By assumption, $L(\,\cdot\,, \lambda)$ is proper, closed, and convex. Therefore,

$$L(\,\cdot\,, \lambda) = \widehat{-f^*(\,\cdot\,, \lambda)} = L^*(\,\cdot\,, \lambda),$$

where the second equality follows from the definition of L^*. □

D. Saddle Points

Strong duality is equivalent to the equality of the saddle points of L and L^*, the Lagrangians defined in Definition 6.22.

6.25 THEOREM. Strong duality, 6.6, holds if and only if

$$\inf_x \sup_\lambda L^*(x, \lambda) = \sup_\lambda \inf_x L(x, \lambda).$$

PROOF. From Theorem 6.23, we have that

$$\begin{aligned}
\inf_x f(x, 0) &= \inf_x \sup_\lambda L^*(x, \lambda), \\
\sup_\lambda f^*(0, \lambda) &= \sup_\lambda \inf_x L(x, \lambda).
\end{aligned}$$

Strong duality means that the expressions on the left are equal. □

Examples

6.26 **Linear programming.** Consider the linear programming problem of minimizing a linear function $c \cdot x$ subject to a collection of m linear inequalities $Ax \geq b$ and positivity of the n variables $x \geq 0$:

$$\begin{aligned}
\text{minimize} \quad & c \cdot x \\
\text{subject to} \quad & Ax \geq b \\
& x \geq 0.
\end{aligned}$$

The m-vector b is called the "right-hand side." Perturbing the right-hand side by y, we get the following primal function

$$f(x, y) = \begin{cases} c \cdot x & x \geq 0 \text{ and } Ax \geq b + y, \\ \infty & \text{otherwise.} \end{cases}$$

Recall the definition of the Lagrangian:

$$L(x, \lambda) = \inf_y \left(-\lambda \cdot y + f(x, y) \right).$$

If $x \not\geq 0$, then the Lagrangian is plus infinity. So, let us suppose that $x \geq 0$. To keep things finite, we need $y \leq Ax - b$. Each component y_i of y will be set either to its largest possible value $(Ax - b)_i$ or will tend to $-\infty$ depending on the sign of its coefficient $-\lambda_i$. If the coefficient is positive, then $-y_i \lambda_i$ is minimized by making y_i as small as possible; that is, by letting it approach $-\infty$. If on the other hand the coefficient is negative, then the term is minimized by setting y_i to its largest possible value $(Ax - b)_i$. If any term goes to minus infinity, then the overall minimum is minus infinity. Hence, the Lagrangian is finite if and only if $-\lambda \leq 0$. To summarize, we have

$$L(x, \lambda) = \begin{cases} \infty & x \not\geq 0, \\ -\infty & x \geq 0, \ \lambda \not\geq 0, \\ -\lambda \cdot (Ax - b) + c \cdot x & x \geq 0, \ \lambda \geq 0. \end{cases}$$

Similar considerations allow us to compute f^* from L:

$$\begin{aligned}
f^*(\xi, \lambda) &= \inf_x \left(-\xi \cdot x + L(x, \lambda) \right) \\
&= \begin{cases} \inf_{x : x \geq 0} \left((c - A^{\mathrm{T}}\lambda - \xi) \cdot x \right) + b \cdot \lambda & \text{if } \lambda \geq 0, \\ -\infty & \text{otherwise} \end{cases} \\
&= \begin{cases} b \cdot \lambda & \text{if } \lambda \geq 0 \text{ and } A^{\mathrm{T}}\lambda \leq c - \xi, \\ -\infty & \text{otherwise.} \end{cases}
\end{aligned}$$

Hence, the dual problem is given by

$$\begin{aligned}
\text{maximize} \quad & b \cdot \lambda \\
\text{subject to} \quad & A^{\mathrm{T}}\lambda \leq c \\
& \lambda \geq 0.
\end{aligned}$$

Reversing directions, it is easy to compute L^* from f^*:

$$L^*(x, \lambda) = \begin{cases} -\infty & \lambda \not\geq 0, \\ \infty & \lambda \geq 0, \ x \not\geq 0, \\ (c - A^{\mathrm{T}}\lambda) \cdot x + b \cdot \lambda & \lambda \geq 0, \ x \geq 0. \end{cases}$$

Note that $L \neq L^*$ in the third quadrant. Note also that f is feasible if and only if

$$\{x \, : \, Ax \geq b, \; x \geq 0\} \neq \emptyset$$

and that f^* is feasible if and only if

$$\{\lambda \, : \, A^{\mathrm{T}}\lambda \leq c, \; \lambda \geq 0\} \neq \emptyset.$$

6.27   *An unbounded LP.* Consider the following trivial linear programming problem:

$$\begin{array}{ll} \text{minimize} & -x \\ \text{subject to} & x \geq 1 \\ & x \geq 0. \end{array}$$

Here, $x \in \mathbb{R}$. The problem is unbounded. The primal function is

$$f(x, y) = \begin{cases} -x & \text{if } x \geq 0 \text{ and } x \geq 1 + y, \\ \infty & \text{otherwise.} \end{cases}$$

The associated dual function is

$$f^*(\xi, \lambda) = \begin{cases} \lambda & \text{if } \lambda \geq 0 \text{ and } \lambda \leq -1 - \xi, \\ -\infty & \text{otherwise,} \end{cases}$$

and the dual problem therefore is

$$\begin{array}{ll} \text{maximize} & \lambda \\ \text{subject to} & \lambda \leq -1 \\ & \lambda \geq 0. \end{array}$$

The dual problem is clearly infeasible. The functions φ and φ^* are easy to compute:

$$\varphi(y) = \inf_x f(x, y) = -\infty$$

and

$$\varphi^*(\xi) = \sup_\lambda f^*(\xi, \lambda) = \begin{cases} -1 - \xi & \xi \leq -1, \\ -\infty & \xi > -1. \end{cases}$$

We see now that f is feasible and φ^* is closed. Therefore the first part of the strong duality theorem tells us that

$$\inf_x f(x, 0) = \sup_\lambda f^*(0, \lambda).$$

Indeed, this is the case. Both sides are minus infinity.

6.28   **Quadratic programming.** The problem of minimizing a convex quadratic function subject to a collection of m linear inequalities $Ax \geq b$ is called the *quadratic programming problem*:

$$\begin{array}{ll} \text{minimize} & c \cdot x + \frac{1}{2} x \cdot Qx \\ \text{subject to} & Ax \geq b. \end{array}$$

As before, we perturb the right-hand side by y to get

$$f(x, y) = \begin{cases} c \cdot x + \frac{1}{2}x \cdot Qx & Ax \geq b + y, \\ \infty & \text{otherwise.} \end{cases}$$

The Lagrangian is easy to compute:

$$\begin{aligned} L(x, \lambda) &= \inf_y \left(-\lambda \cdot y + f(x, y) \right) \\ &= \begin{cases} -\lambda \cdot (Ax - b) + c \cdot x + \frac{1}{2}x \cdot Qx & \lambda \geq 0, \\ -\infty & \text{otherwise.} \end{cases} \end{aligned}$$

Computing f^* from L is a little trickier than before. We start out in the same way:

$$\begin{aligned} f^*(\xi, \lambda) &= \inf_x \left(-\xi \cdot x + L(x, \lambda) \right) \\ &= \begin{cases} b \cdot \lambda + \inf_x \left((c - \xi - A^{\mathrm{T}}\lambda) \cdot x + \frac{1}{2}x \cdot Qx \right) & \lambda \geq 0, \\ -\infty & \text{otherwise.} \end{cases} \end{aligned}$$

The function $x \mapsto (c - \xi - A^{\mathrm{T}}\lambda) \cdot x + \frac{1}{2}x \cdot Qx$ is a convex quadratic function and it is being minimized over all x in $\mathbb{R}^n$. Hence, its minimum is found by taking the gradient of the function and setting that to zero. That is, the expression is minimized at the point that solves

$$c - \xi - A^{\mathrm{T}}\lambda + Qx = 0.$$

Therefore, the minimum value is

$$\begin{aligned} (c - \xi - A^{\mathrm{T}}\lambda) \cdot x + \frac{1}{2}x \cdot Qx &= (c - \xi - A^{\mathrm{T}}\lambda + Qx) \cdot x - \frac{1}{2}x \cdot Qx \\ &= -\frac{1}{2}x \cdot Qx, \end{aligned}$$

and the dual function f^* is given by

$$f^*(\xi, \lambda) = \begin{cases} b \cdot \lambda - \frac{1}{2}x \cdot Qx \big|_{x:c-\xi-A^{\mathrm{T}}\lambda+Qx=0} & \lambda \geq 0, \\ -\infty & \text{otherwise.} \end{cases}$$

Hence, the dual problem is given by

$$\begin{aligned} \text{maximize} \quad & b \cdot \lambda - \frac{1}{2}x \cdot Qx \\ \text{subject to} \quad & A^{\mathrm{T}}\lambda - Qx = c \\ & \lambda \geq 0. \end{aligned}$$

Exercises

6.29   Let $f \colon \mathbb{R} \mapsto \mathbb{R}$ be a proper, closed, convex function and suppose that it is differentiable at those points where it is finite. Show that $\widehat{f}$ is differentiable on the set where it is finite and

$$\widehat{f}\,' = (f')^{-1}.$$

6.30   Consider the linear programming problem: maximize x subject to $0 \cdot x \leq -1$, $x \geq 0$. Compute L, f, and f^*. Does strong duality hold?

6.31   Compute the dual of the linear programming problem: minimize $c \cdot x$ subject to $Ax = b$, $x \geq 0$. Hint: use a perturbation to the right-hand side by replacing b with $b + y$.

6.32   Given a convex function $f \colon \mathbb{R}^n \mapsto \mathbb{R}$ and a convex and monotonically increasing function $g \colon \mathbb{R} \mapsto \mathbb{R} \cup \{\infty\}$, show that the composition $g \circ f$ is convex.

6.33   Show that the geometric mean

$$f(x_1, \ldots, x_n) = \begin{cases} (x_1 \cdots x_n)^{1/n} & \text{if } x_i \geq 0 \text{ for all } i, \\ -\infty & \text{otherwise.} \end{cases}$$

is concave.

6.34   Compute f^* when

$$f(x, y) = \begin{cases} x & \text{if } x^2 \leq y, \\ \infty & \text{if } x^2 > y. \end{cases}$$

Does strong duality hold? Is the primal optimal solution attained? Is the dual optimal solution attained?

6.35   Compute f^* when

$$f(x, y) = \begin{cases} e^{-\sqrt{xy}} & \text{if } x, y \geq 0, \\ \infty & \text{otherwise.} \end{cases}$$

CHAPTER 7

Measure and Integration

This chapter is an introduction to measure and integration on abstract spaces. The treatment is driven by the needs of modern analysis and probability. The result is a robust concept of integration that extends the concepts familiar in calculus.

A. Algebras

Let E be a set. Let $\mathcal{E}$ be a collection of subsets of E; see Chap. 1 for the notations and terminology. The collection $\mathcal{E}$ is called an *algebra* on E if it includes E and is closed under the operations of complementation and finite unions. It is called a *σ-algebra* if it is an algebra that is closed under countable unions. To reiterate, $\mathcal{E}$ is a σ-algebra on E if

7.1
(i) $E \in \mathcal{E}$,
(ii) $A \in \mathcal{E} \Rightarrow A^c \in \mathcal{E}$,
(iii) $A_1, A_2, \ldots \in \mathcal{E} \Rightarrow \bigcup_{n=1}^{\infty} A_n \in \mathcal{E}$.

A σ-algebra is also closed under countable intersections because the intersection of a collection of subsets is the complement of the union of the complements of those subsets.

The sets E and $\emptyset$ belong to every σ-algebra on E. Thus, the simplest σ-algebra on E is $\mathcal{E} = \{\emptyset, E\}$; it is called the trivial σ-algebra. The largest is the collection of all subsets, denoted by 2^E or $\mathcal{P}(E)$; it is called the discrete σ-algebra on E.

The intersection of an arbitrary family (countable or uncountable) of σ-algebras on E is again a σ-algebra. If $\mathcal{C}$ is a collection of subsets of E, the intersection of all σ-algebras containing $\mathcal{C}$ is the smallest σ-algebra that contains $\mathcal{C}$; it is called the σ-algebra *generated* by $\mathcal{C}$ and is denoted by $\sigma(\mathcal{C})$.

If E is a metric space, then the σ-algebra generated by the collection of all open subsets is called the *Borel σ-algebra* on E; it is denoted by $\mathcal{B}(E)$, and its elements are called Borel sets. Thus, every open set, every closed set, and every set obtained from open and closed sets through countably many set operations are all Borel sets.

E. Çınlar and R.J. Vanderbei, *Real and Convex Analysis*, Undergraduate Texts in Mathematics,   115
DOI 10.1007/978-1-4614-5257-7_7, © Springer Science+Business Media New York 2013

Monotone Class Theorem

   This useful theorem simplifies the task of showing that a given collection is a σ-algebra. Throughout this subsection, E is an arbitrary set.

   A collection $\mathcal{C}$ of subsets of E is called a π-*system* if it is closed under finite intersections, that is, if

7.2 $$A, B \in \mathcal{C} \Rightarrow A \cap B \in \mathcal{C}.$$

A collection $\mathcal{D}$ of subsets of E is called a d-*system* on E if

7.3
   (i) $E \in \mathcal{D}$,
   (ii) $A, B \in \mathcal{D}$ and $B \subset A \Rightarrow A \setminus B \in \mathcal{D}$,
   (iii) $(A_n) \subset \mathcal{D}$ and $A_n \nearrow A \Rightarrow A \in \mathcal{D}$.

On the last line, we wrote $(A_n) \subset \mathcal{D}$ to mean that (A_n) is a sequence of elements of $\mathcal{D}$, and we wrote $A_n \nearrow A$ to mean that $A_1 \subset A_2 \subset \cdots$ and $\bigcup_n A_n = A$.

7.4 PROPOSITION. Let $\mathcal{E}$ be a collection of subsets of E. Then, $\mathcal{E}$ is a σ-algebra on E if and only if $\mathcal{E}$ is both a π-system and a d-system on E.

PROOF. If $\mathcal{E}$ is a σ-algebra then it is obviously a π-system and a d-system. To show the converse, suppose that $\mathcal{E}$ is both a π-system and a d-system. Now, 7.3 (i) and 7.3 (ii) show that $\mathcal{E}$ is closed under complements. Because $A \cup B = (A^c \cap B^c)^c$, this implies that $\mathcal{E}$ is closed under unions: if $A, B \in \mathcal{E}$ then $A^c, B^c \in \mathcal{E}$, and thus $A^c \cap B^c \in \mathcal{E}$ because $\mathcal{E}$ is a π-system, and hence $(A^c \cap B^c)^c \in \mathcal{E}$. This implies that $\mathcal{E}$ is closed under countable unions. If $A_1, A_2, \ldots \in \mathcal{E}$, put

$$B_1 = A_1, \quad B_2 = B_1 \cup A_2, \quad B_3 = B_2 \cup A_3, \ldots.$$

Each B_n belongs to $\mathcal{E}$ by what we have just shown. Obviously, $B_1 \subset B_2 \subset \cdots$ and $\bigcup_n B_n = \bigcup_n A_n$. Thus, using the property 7.3 (iii) of the d-system $\mathcal{E}$, we see that $\bigcup_n A_n \in \mathcal{E}$. □

   The following lemma is needed in the proof of the main theorem. Its proof is obtained by checking the conditions of 7.3 one by one; we leave it as an exercise.

7.5 LEMMA. Let $\mathcal{D}$ be a d-system on E. Fix D in $\mathcal{D}$ and let

$$\hat{\mathcal{D}} = \{A \in \mathcal{D} : A \cap D \in \mathcal{D}\}.$$

Then, $\hat{\mathcal{D}}$ is again a d-system.

The following is the main result of this section. It is called Dynkin's monotone class theorem.

7.6 THEOREM. If a d-system contains a π-system, then it contains also the σ-algebra generated by that π-system.

PROOF. Let C be a π-system. Let D be the smallest d-system on E that contains C. We need to show that $D \supset \sigma(C)$. To that end, because $\sigma(C)$ is the smallest σ-algebra containing C, it is sufficient to show that D is a σ-algebra. For this, it is in turn sufficient to show that D is a π-system (and then Proposition 7.4 implies that the d-system D is a σ-algebra).

Fix B in C and let $D_1 = \{A \in D : A \cap B \in D\}$. Because $B \in C \subset D$, Lemma 7.5 shows that D_1 is a d-system. Moreover, $D_1 \supset C$ since $A \cap B \in C \subset D$ for every $A \in C$ by the fact that C is a π-system. So D_1 must contain the smallest d-system containing C, that is, $D_1 \supset D$. In other words, $A \cap B \in D$ for every A in D and B in C.

Next, fix A in D and let $D_2 = \{B \in D : A \cap B \in D\}$. We have just shown that $D_2 \supset C$. Moreover, by Lemma 7.5 again, D_2 is a d-system. Thus, $D_2 \supset D$. In other words, $A \cap B \in D$ for every A in D and B in D, that is, D is a π-system. □

Exercises

7.7 *Partitions.* A partition of E is a countable disjointed collection of subsets whose union is E. It is called a finite partition if it has only finitely many elements.

(a) Let $\{A, B, C\}$ be a partition of E. List the elements of the σ-algebra generated by this partition.

(b) Let C be a partition of E. Let $\mathcal{E}$ be the collection of all countable unions of elements of C. Show that $\mathcal{E}$ is a σ-algebra. Show that, in fact, $\mathcal{E} = \sigma(C)$.

Generally, if C is not a partition, the elements of $\sigma(C)$ cannot be obtained through such explicit constructions.

7.8 Let $\mathcal{B}$ and C be two collections of subsets of E. If $\mathcal{B} \subset C$, then $\sigma(\mathcal{B}) \subset \sigma(C)$. If $\mathcal{B} \subset \sigma(C) \subset \sigma(\mathcal{B})$, then $\sigma(\mathcal{B}) = \sigma(C)$. Show these.

7.9 *Borel σ-algebra on* $\mathbb{R}$. Show that $\mathcal{B}(\mathbb{R})$ is generated by the collection of all open intervals. Hint: Recall that every open subset of $\mathbb{R}$ is a countable union of open intervals.

7.10 *Continuation.* Show that every interval of $\mathbb{R}$ is a Borel set. In particular, $(-\infty, x)$, $(-\infty, x]$, $(x, y]$, $[x, y]$ are all Borel sets. Every singleton $\{x\}$ is a Borel set.

7.11 Show that $\mathcal{B}(\mathbb{R})$ is also generated by any one of the following:

  (a) The collection of all intervals of the form (x, ∞)
  (b) The collection of all intervals of the form $(x, y]$
  (c) The collection of all intervals of the form $[x, y]$
  (d) The collection of all intervals of the form $(-\infty, x]$
  (e) The collection of all intervals of the form (x, ∞) with x rational

B. Measurable Spaces and Functions

A *measurable space* is a pair $(E, \mathcal{E})$, where E is a set and $\mathcal{E}$ is a σ-algebra on E. Then, the elements of $\mathcal{E}$ are called *measurable sets*. When E is a metric space and $\mathcal{E} = \mathcal{B}(E)$, the Borel σ-algebra on E, the measurable sets are also called *Borel sets*.

Let $(E, \mathcal{E})$ and $(F, \mathcal{F})$ be measurable spaces and let f be a mapping from E into F. Then, f is said to be *measurable* relative to $\mathcal{E}$ and $\mathcal{F}$ if $f^{-1}B \in \mathcal{E}$ for every B in $\mathcal{F}$ (these are the functions we wish to be able to integrate). Recall that, for subsets B of F, $f^{-1}B$ denotes the inverse image of B under f; see 1.5 and Exercise 1.6. If E and F are metric spaces and $\mathcal{E} = \mathcal{B}(E)$ and $\mathcal{F} = \mathcal{B}(F)$ and $f: E \mapsto F$ is measurable relative to $\mathcal{E}$ and $\mathcal{F}$, then f is also called a *Borel function*.

Measurable Functions

The following proposition uses Exercise 1.6 to reduce the checks for measurability.

7.12 PROPOSITION. Let $(E, \mathcal{E})$ and $(F, \mathcal{F})$ be measurable spaces. For $f: E \mapsto F$ to be measurable relative to $\mathcal{E}$ and $\mathcal{F}$, it is necessary and sufficient that $f^{-1}B \in \mathcal{E}$ for every B in $\mathcal{F}_0$ for some collection $\mathcal{F}_0$ that generates $\mathcal{F}$.

PROOF. The necessity part is trivial. To prove the sufficiency, let $\mathcal{F}_0 \subset \mathcal{F}$ be such that $\sigma(\mathcal{F}_0) = \mathcal{F}$ and suppose that $f^{-1}B \in \mathcal{E}$ for every B in $\mathcal{F}_0$. We need to show that, then,

$$\mathcal{F}_1 = \{B \in \mathcal{F} : f^{-1}B \in \mathcal{E}\}$$

is equal to $\mathcal{F}$. For this, it is sufficient to show that $\mathcal{F}_1$ is a σ-algebra because $\mathcal{F}_1 \supset \mathcal{F}_0$ by hypothesis and $\mathcal{F}$ is the smallest σ-algebra containing $\mathcal{F}_0$. But checking that $\mathcal{F}_1$ is a σ-algebra is easy in view of the relations given in Exercise 1.6. $\qquad\square$

Borel Functions

Let E and F be metric spaces and let $\mathcal{E}$ and $\mathcal{F}$ be their respective Borel σ-algebras. Let $f \colon E \mapsto F$. Since $\mathcal{F}$ is generated by the open subsets of F, in order for f to be a Borel function, it is necessary and sufficient that $f^{-1}B \in \mathcal{E}$ for every open subset B of F; this is an immediate corollary of the preceding proposition. In particular, if f is continuous, then $f^{-1}B$ is open in E for every open $B \subset F$. Thus, every continuous function $f \colon E \mapsto F$ is Borel measurable. The converse is generally false.

Compositions of Functions

Let $(E, \mathcal{E})$, $(F, \mathcal{F})$, and $(G, \mathcal{G})$ be measurable spaces. Let $f \colon E \mapsto F$ and $g \colon F \mapsto G$. Then, their composition $g \circ f \colon x \mapsto g(f(x))$ is a mapping from E into G. The following proposition will be recalled by the phrase "measurable functions of measurable functions are measurable."

7.13 PROPOSITION. If f is measurable relative to $\mathcal{E}$ and $\mathcal{F}$, and if g is measurable relative to $\mathcal{F}$ and $\mathcal{G}$, then $g \circ f$ is measurable relative to $\mathcal{E}$ and $\mathcal{G}$.

PROOF. Recall that $(g \circ f)^{-1}C = f^{-1}(g^{-1}C)$ for every $C \subset G$. If $C \in \mathcal{G}$ and g is measurable, then $B = g^{-1}C$ is in $\mathcal{F}$. Therefore, if f is measurable, $f^{-1}B = f^{-1}(g^{-1}C)$ is in $\mathcal{E}$ for every C in $\mathcal{G}$. $\qquad\square$

Numerical Functions

By a *numerical function* on E we mean a mapping from E into $\bar{\mathbb{R}}$ or some subset thereof. Such a function is said to be *positive* if all its values are in $\bar{\mathbb{R}}_+$ and is said to be real-valued if all its values are in $\mathbb{R}$. If $(E, \mathcal{E})$ is a measurable space and f is a numerical function on E, then f is said to be $\mathcal{E}$-*measurable* if it is measurable with respect to $\mathcal{E}$ and $\mathcal{B}(\bar{\mathbb{R}})$.

Let $(E, \mathcal{E})$ be a measurable space and let f be a numerical function on E. Using Proposition 7.12 with $F = \bar{\mathbb{R}}$ and $\mathcal{F} = \mathcal{B}(\bar{\mathbb{R}})$ and recalling Exercise 7.11, we see that the following holds.

7.14 PROPOSITION. The numerical function f is $\mathcal{E}$-measurable if and only if any one of the following is true:

    (a) $\{x \in E : f(x) \le r\} \in \mathcal{E}$ for every r in $\mathbb{R}$,
    (b) $\{x \in E : f(x) > r\} \in \mathcal{E}$ for every r in $\mathbb{R}$,
    (c) $\{x \in E : f(x) < r\} \in \mathcal{E}$ for every r in $\mathbb{R}$, etc.

7.15 COROLLARY. Suppose that $f : E \mapsto F$, where F is a countable subset of $\bar{\mathbb{R}}$. Then, f is $\mathcal{E}$-measurable if and only if $\{x : f(x) = a\} \in \mathcal{E}$ for every a in F.

PROOF. Necessity is trivial since each singleton $\{a\}$ is a Borel set. For the sufficiency, fix r in $\mathbb{R}$ and note that $\{x : f(x) \le r\}$ is the union of $\{x : f(x) = a\}$ over all $a \le r$, $a \in F$, and therefore belongs to $\mathcal{E}$ since it is a countable union of the sets $\{x : f(x) = a\} \in \mathcal{E}$. Thus, f is $\mathcal{E}$-measurable by the preceding proposition.   □

Positive and Negative Parts of a Function

    Let $(E, \mathcal{E})$ be a measurable space. Let f be a numerical function on E. Then,[1]

$$f^+ = f \vee 0, \quad f^- = -(f \wedge 0)$$

are called the positive part of f and negative part of f, respectively. Note that both f^+ and f^- are positive functions and

$$f = f^+ - f^-.$$

7.16 PROPOSITION. The function f is $\mathcal{E}$-measurable if and only if both f^+ and f^- are $\mathcal{E}$-measurable.

    The proof is left as an exercise. The decomposition $f = f^+ - f^-$ enables us to state most results for positive functions only, since it is easy to obtain the corresponding result for arbitrary f.

Indicators and Simple Functions

    Let $A \subset E$. Its *indicator*, denoted by 1_A, is defined by

$$1_A(x) = \begin{cases} 1 & \text{if } x \in A, \\ 0 & \text{if } x \notin A. \end{cases}$$

Obviously, 1_A is $\mathcal{E}$-measurable if and only if $A \in \mathcal{E}$.

---

[1]For $a, b \in \bar{\mathbb{R}}$ we write $a \vee b$ for the maximum of a and b, and $a \wedge b$ for the minimum. The notation extends to functions: $f \vee g$ is the function whose value at x is $f(x) \vee g(x)$; similarly for $f \wedge g$.

A function f on E is said to be *simple* if it has the form

7.17
$$f = \sum_{1}^{n} a_i 1_{A_i}$$

for some integer n, real numbers $a_1, \ldots, a_n$, and measurable sets $A_1, \ldots, A_n$. It is clear that, then, there exist an integer $m \geq 1$, distinct real numbers $b_1, \ldots, b_m$, and a measurable partition $\{B_1, \ldots, B_m\}$ of E such that $f = \sum_{1}^{m} b_i 1_{B_i}$; this latter representation is called the *canonical form* of the simple function f.

Every simple function on E is $\mathcal{E}$-measurable; this is immediate from Corollary 7.15 applied to the canonical form of f. Conversely, if f is $\mathcal{E}$-measurable and f takes only finitely many values, and all those values are real, then f is simple.

In particular, every constant is a simple function. Moreover, if f and g are simple, then so are

$$f + g, \quad f - g, \quad fg, \quad f/g, \quad f \vee g, \quad f \wedge g,$$

except that in the case of f/g one must make sure that g is never 0.

Approximations by Simple Functions

We start by constructing a sequence of simple functions that approximate the identity function d from $\bar{\mathbb{R}}_+$ into $\bar{\mathbb{R}}_+$. For each $n \in \mathbb{N}^*$, let

7.18
$$d_n(x) = \begin{cases} k/2^n & \text{if } \frac{k}{2^n} \leq x < \frac{k+1}{2^n}, \quad k \in \{0, 1, \ldots, n2^n - 1\}, \\ n & \text{if } x \geq n. \end{cases}$$

The following lemma should be self-evident.

7.19 LEMMA. Each d_n is a simple Borel function on $\bar{\mathbb{R}}_+$. Each d_n is right-continuous and increasing. The sequence (d_n) is increasing pointwise to the function $d \colon x \mapsto x$.

The following theorem characterizes all $\mathcal{E}$-measurable positive functions and, via Proposition 7.16, all $\mathcal{E}$-measurable functions.

7.20 THEOREM. A positive function on E is $\mathcal{E}$-measurable if and only if it is the limit of an increasing sequence of simple positive functions.

PROOF. *Necessity.* Let $f \colon E \mapsto \bar{\mathbb{R}}_+$ be $\mathcal{E}$-measurable. Let the functions d_n be defined by 7.18. Since each d_n is a measurable function from $\bar{\mathbb{R}}_+$ into $\bar{\mathbb{R}}_+$, and since measurable functions of measurable functions are measurable, the function $f_n = d_n \circ f$ is $\mathcal{E}$-measurable for each n. Since d_n is simple, so is f_n. Finally, $\lim f_n(x) = \lim d_n(f(x)) = f(x)$ since $\lim d_n(y) = y$ for all y in $\bar{\mathbb{R}}_+$. Thus, f is the limit of the sequence (f_n) of simple positive functions and $f_1 \leq f_2 \leq \cdots$ since $d_1 \leq d_2 \leq \cdots$.

Sufficiency. Let $f_1 \leq f_2 \leq \cdots$ be simple and positive and let $f = \lim f_n$. Now, for each x in E and r in $\mathbb{R}$, we have $f(x) \leq r$ if and only if $f_n(x) \leq r$ for all n; thus,

$$\{x \in E : f(x) \leq r\} = \bigcap_{n=1}^{\infty} \{x \in E : f_n(x) \leq r\}$$

for each r in $\mathbb{R}$. Since the f_n are simple (and therefore measurable), each factor on the right side belongs to $\mathcal{E}$ and, therefore, so does the intersection. Hence, f is $\mathcal{E}$-measurable by Proposition 7.14. □

Limits of Sequences of Functions

Let $(E, \mathcal{E})$ be a measurable space and let (f_n) be a sequence of numerical functions on E.

7.21 THEOREM. Suppose that each f_n is $\mathcal{E}$-measurable. Then, each one of

$$\inf f_n, \quad \sup f_n, \quad \liminf f_n, \quad \limsup f_n$$

is again $\mathcal{E}$-measurable. Moreover, if $\lim f_n$ exists, then it is $\mathcal{E}$-measurable.

PROOF. For x in E and r in $\mathbb{R}$, we have $\inf f_n(x) \geq r$ if and only if $f_n(x) \geq r$ for all n. Thus, for each r in $\mathbb{R}$,

$$\{x \in E : \inf f_n(x) \geq r\} = \bigcap_n \{x \in E : f_n(x) \geq r\}.$$

Now, $\{x : f_n(x) \geq r\} \in \mathcal{E}$ for each n by the measurability of f_n, and therefore the intersection on the right side belongs to $\mathcal{E}$ since $\mathcal{E}$ is closed under countable intersections. Thus, $\inf f_n$ is $\mathcal{E}$-measurable by Proposition 7.14.

The proof that $\sup f_n$ is $\mathcal{E}$-measurable follows via similar reasoning upon noting that

$$\{x \in E : \sup f_n(x) \leq r\} = \bigcap_n \{x \in E : f_n(x) \leq r\}.$$

It follows from these that

$$\liminf f_n = \sup_m \inf_{n \geq m} f_n, \quad \limsup f_n = \inf_m \sup_{n \geq m} f_n$$

are both $\mathcal{E}$-measurable. Finally, $\lim f_n$ exists if and only if $\liminf f_n = \limsup f_n$, and then $\lim f_n$ is the common limit; so, it must be $\mathcal{E}$-measurable. □

Monotone Classes of Functions

Often we are interested in showing that a certain property holds for all measurable functions. The following are useful in such quests.

Let $\mathcal{M}$ be a collection of functions on E. Then, $\mathcal{M}$ is called a *monotone class of functions* provided that

7.22
$$\begin{array}{ll} \text{(i)} & 1 \in \mathcal{M}, \\ \text{(ii)} & f, g \in \mathcal{M}, \text{ and } a, b \in \mathbb{R} \Rightarrow af + bg \in \mathcal{M}, \\ \text{(iii)} & (f_n) \subset \mathcal{M}, f_n \geq 0, \text{ and } f_n \nearrow f \Rightarrow f \in \mathcal{M}. \end{array}$$

The following is called the monotone class theorem for functions.

7.23 THEOREM. Let $\mathcal{M}$ be a monotone class of functions on E. Suppose that $1_A \in \mathcal{M}$ for every A in $\mathcal{C}$ for some π-system $\mathcal{C}$ that generates the σ-algebra $\mathcal{E}$. Then, $\mathcal{M}$ includes all positive $\mathcal{E}$-measurable functions.

PROOF. We start by showing that $1_A \in \mathcal{M}$ for every $A \in \mathcal{E}$. To this end, let

$$\mathcal{D} = \{A \in \mathcal{E} : 1_A \in \mathcal{M}\}.$$

Using the properties 7.22 of $\mathcal{M}$, it is easy to check that $\mathcal{D}$ is a d-system. Moreover, $\mathcal{D} \supset \mathcal{C}$ by hypothesis. Thus, by Dynkin's monotone class theorem, $\mathcal{D} \supset \sigma(\mathcal{C}) = \mathcal{E}$. In other words, $1_A \in \mathcal{M}$ for every $A \in \mathcal{E}$.

Consequently, in view of property 7.22(ii), $\mathcal{M}$ includes all simple positive $\mathcal{E}$-measurable functions.

Let f be a positive $\mathcal{E}$-measurable function. By Theorem 7.20, there exists a sequence of positive simple functions $f_n \nearrow f$. Since each f_n is in $\mathcal{M}$ by the preceding step, 7.22(iii) implies that f is in $\mathcal{M}$.

$\square$

Notation

We shall write $f \in \mathcal{E}$ to mean that f is an $\mathcal{E}$-measurable function. Thus, $\mathcal{E}$ stands both for a σ-algebra and for the collection of all numerical functions measurable with respect to it. Furthermore, we shall use the notation

$$\mathcal{F}_+ = \{f \in \mathcal{F} : f \geq 0\}$$

for any collection $\mathcal{F}$ of numerical functions. Thus, in particular, $\mathcal{E}_+$ is the collection of all positive $\mathcal{E}$-measurable functions.

Exercises

7.24 *Trace spaces.* Let $(E, \mathcal{E})$ be a measurable space and let $D \subset E$ be fixed. Show that

$$\mathcal{D} = \{A \cap D : A \in \mathcal{E}\}$$

is a σ-algebra on $\mathcal{D}$. Then, $\mathcal{D}$ is called the trace of $\mathcal{E}$ on D, and $(D, \mathcal{D})$ is called the trace of $(E, \mathcal{E})$ on D.

7.25 *σ-Algebra generated by a function.* Let E be a set and let $(F, \mathcal{F})$ be a measurable space. Let f be a mapping from E into F and set

$$f^{-1}(\mathcal{F}) = \{f^{-1}B : B \in \mathcal{F}\}.$$

Use Exercise 1.6 to show that $f^{-1}(\mathcal{F})$ is a σ-algebra on E; it is called the σ-algebra on E generated by f. It is the smallest σ-algebra on E such that f is measurable relative to it and $\mathcal{F}$.

7.26 *Product spaces.* Let $(E, \mathcal{E})$ and $(F, \mathcal{F})$ be measurable spaces. A rectangle $A \times B$ is said to be measurable if $A \in \mathcal{E}$ and $B \in \mathcal{F}$. Show that the collection of all measurable rectangles form a π-system. The σ-algebra on $E \times F$ generated by that π-system is denoted by $\mathcal{E} \otimes \mathcal{F}$ and is called the product σ-algebra. Further, $(E \times F, \mathcal{E} \otimes \mathcal{F})$ is called the product of $(E, \mathcal{E})$ and $(F, \mathcal{F})$ and is denoted by $(E, \mathcal{E}) \times (F, \mathcal{F})$ also. If $(E, \mathcal{E}) = (F, \mathcal{F})$, then it is usual to write E^2 for $E \times F$ and $\mathcal{E}^2$ for $\mathcal{E} \otimes \mathcal{F}$. In particular, it can be shown that $(\mathbb{R}^2, \mathcal{B}(\mathbb{R}^2)) = (\mathbb{R}, \mathcal{B}(\mathbb{R})) \times (\mathbb{R}, \mathcal{B}(\mathbb{R}))$, and by an obvious extension, $(\mathbb{R}^n, \mathcal{B}(\mathbb{R}^n)) = (\mathbb{R}, \mathcal{B}(\mathbb{R})) \times \cdots \times (\mathbb{R}, \mathcal{B}(\mathbb{R}))$, n times.

7.27 *Continuation.* Let $(E, \mathcal{E})$, $(F, \mathcal{F})$, $(G, \mathcal{G})$ be measurable spaces. Let $f \colon E \mapsto F$ be measurable relative to $\mathcal{E}$ and $\mathcal{F}$, and let $g \colon E \mapsto G$ be measurable relative to $\mathcal{E}$ and $\mathcal{G}$. Then,

$$h(x) = (f(x), g(x)), \quad x \in E,$$

defines a mapping from E into $F \times G$. Show that h is measurable relative to $\mathcal{E}$ and $\mathcal{F} \otimes \mathcal{G}$.

In particular, a function $f \colon E \mapsto \mathbb{R}^n$ is measurable relative to $\mathcal{E}$ and $\mathcal{B}(\mathbb{R}^n)$ if and only if its coordinates are measurable relative to $\mathcal{E}$ and $\mathcal{B}(\mathbb{R})$; recall that the coordinates of f are the functions $f_1, \ldots, f_n$ such that $f(x) = (f_1(x), \ldots, f_n(x))$, $x \in E$.

7.28 *Discrete spaces.* A measurable space $(E, \mathcal{E})$ is said to be *discrete* if E is countable and $\mathcal{E}$ is the σ-algebra of all subsets of E. Then, show that every numerical function on E is $\mathcal{E}$-measurable.

7.29 Suppose that $\mathcal{E}$ is generated by a partition of E. Show that, then, a numerical function on E is $\mathcal{E}$-measurable if and only if it is constant over each member of that partition.

7.30 *Approximation by simple functions.* Show that a numerical function on E is $\mathcal{E}$-measurable if and only if it is the limit of a sequence of simple functions.

7.31 *Arithmetic operations.* Let f and g be $\mathcal{E}$-measurable. Show that, then, each one of

$$f + g, \quad f - g, \quad f \cdot g, \quad f/g, \quad f \vee g, \quad f \wedge g$$

is $\mathcal{E}$-measurable provided that it is well-defined.

7.32 *Functions on $\mathbb{R}$.* Let $f: \mathbb{R} \mapsto \mathbb{R}_+$ be increasing. Show that it is a Borel function.

7.33 *Step functions.* A function $f: \mathbb{R}_+ \mapsto \mathbb{R}$ is called a step function if there exists a subdivision

$$0 = x_0 < x_1 < x_2 < \cdots$$

with $\lim x_n = +\infty$ such that f is constant over (x_i, x_{i+1}) for each $i \geq 0$. Show that every such f is a Borel function.

7.34 *Right-continuous functions.* Show that every right-continuous function $f: \mathbb{R}_+ \mapsto \mathbb{R}$ is Borel measurable. Similarly, every left-continuous function is Borel. Hint for right-continuous f: define $d_n(x) = (k + 1)/2^n$ if $k/2^n \leq x < (k + 1)/2^n$ for some $k = 0, 1, 2, \ldots$ for $n = 1, 2, \ldots$. Show that d_n is Borel. Let $f_n(x) = f(d_n(x))$. Show that each f_n is a step function, and show that $f_n \to f$ as $n \to \infty$.

C. Measures

Let $(E, \mathcal{E})$ be a measurable space. A *measure* on $(E, \mathcal{E})$ is a mapping $\mu: \mathcal{E} \mapsto \bar{\mathbb{R}}_+$ such that

(a) $\mu(\emptyset) = 0$,
(b) $\mu(\bigcup_n A_n) = \sum_n \mu(A_n)$ for every disjointed sequence (A_n) in $\mathcal{E}$.

The latter condition is called *countable additivity*.

A *measure space* is a triplet $(E, \mathcal{E}, \mu)$ where E is a set, $\mathcal{E}$ is a σ-algebra on E, and μ is a measure on $(E, \mathcal{E})$.

7.35 PROPOSITION. Let μ be a measure on $(E, \mathcal{E})$. Then, the following hold for all measurable sets A, B, and A_n, $n \geq 1$:

Finite additivity: $A \cap B = \emptyset$ implies that $\mu(A \cup B) = \mu(A) + \mu(B)$.
Monotonicity: $A \subset B$ implies that $\mu(A) \leq \mu(B)$.
Sequential continuity: $A_n \nearrow A$ implies that $\mu(A_n) \nearrow \mu(A)$.
Boole's inequality: $\mu(\bigcup_n A_n) \leq \sum_n \mu(A_n)$.

PROOF. Finite additivity is a particular instance of the countable additivity of μ: take $A_1 = A$, $A_2 = B$, $A_3 = A_4 = \cdots = \emptyset$. Monotonicity follows from it and the positivity of μ: if $A \subset B$,

$$\mu(B) = \mu(A) + \mu(B \setminus A) \geq \mu(A)$$

since $\mu(B \setminus A) \geq 0$. Sequential continuity follows from (and is equivalent to) countable additivity: suppose that $A_n \nearrow A$; then,

$$B_1 = A_1, \quad B_2 = A_2 \setminus A_1, \quad B_3 = A_3 \setminus A_2, \quad \ldots$$

are disjoint, their union is A, and the union of the first n is A_n; hence, the sequence of numbers $\mu(A_n)$ increases by the monotonicity of μ, and

$$\lim \mu(A_n) = \lim \mu \left(\bigcup_1^n B_i \right) = \lim_n \sum_1^n \mu(B_i) = \sum_1^\infty \mu(B_i) = \mu \left(\bigcup_1^\infty B_i \right) = \mu(A).$$

Finally, Boole's inequality follows from the observation that

$$\mu(A \cup B) = \mu(A) + \mu(B \setminus A) \leq \mu(A) + \mu(B).$$

$\square$

Arithmetic of Measures

Let $(E, \mathcal{E})$ be a measurable space. If μ is a measure on it and if $c \geq 0$ is a constant, then $c\mu$ is again a measure. If μ and ν are measures on $(E, \mathcal{E})$, so is $\mu + \nu$. If $\mu_1, \mu_2, \ldots$ are measures, then so is $\mu = \sum \mu_m$: it is obvious that $\mu(\emptyset) = 0$, and if $A_1, A_2, \ldots$ are disjoint then

$$\begin{aligned}
\mu \left(\bigcup_n A_n \right) &= \sum_m \mu_m \left(\bigcup_n A_n \right) \\
&= \sum_m \sum_n \mu_m(A_n) \\
&= \sum_n \sum_m \mu_m(A_n) = \sum_n \mu(A_n),
\end{aligned}$$

where the crucial step (where the order of summation is changed) is justified by the elementary fact that

$$\sum_m \sum_n a_{mn} = \sum_n \sum_m a_{mn}$$

if $a_{mn} \geq 0$ for all m, n.

Finite, σ-Finite, Σ-Finite Measures

Let μ be a measure on $(E, \mathcal{E})$. It is said to be *finite* if $\mu(E) < \infty$. It is called a *probability measure* if $\mu(E) = 1$. It is said to be σ-finite if there exists a measurable partition (E_n) of E such that $\mu(E_n) < \infty$ for each n. It is said to be Σ-finite if there exist finite measures $\mu_1, \mu_2, \ldots$ such that $\mu = \sum \mu_n$. Note that every finite measure is trivially σ-finite, every σ-finite measure is Σ-finite. The converses are false (see Exercise 7.49).

Specification of Measures

It is generally difficult to specify $\mu(A)$ for each A, simply because there are too many A in a σ-algebra. The following proposition is helpful in reducing the task to that of specifying $\mu(A)$ for those A belonging to a π-system that generates the given σ-algebra.

7.36 PROPOSITION. Let μ and ν be measures on $(E, \mathcal{E})$. Suppose that $\mu(E) = \nu(E) < \infty$, and that μ and ν agree on a π-system generating $\mathcal{E}$. Then, $\mu = \nu$.

PROOF. Let $\mathcal{C}$ be a π-system with $\sigma(\mathcal{C}) = \mathcal{E}$. Suppose that $\mu(A) = \nu(A)$ for every A in $\mathcal{C}$. We need to show that, then, $\mu(A) = \nu(A)$ for every A in $\mathcal{E}$. This amounts to showing that

$$\mathcal{D} = \{A \in \mathcal{E} : \mu(A) = \nu(A)\}$$

contains $\mathcal{E}$. Now, $\mathcal{D} \supset \mathcal{C}$ by hypothesis, and it is straightforward to check that $\mathcal{D}$ is a d-system. Thus, by Dynkin's monotone class theorem, $\mathcal{D} \supset \sigma(\mathcal{C}) = \mathcal{E}$. □

7.37 COROLLARY. Let μ and ν be probability measures on $(\mathbb{R}, \mathcal{B}(\mathbb{R}))$. Then, $\mu = \nu$ if and only if, for every x in $\mathbb{R}$,

$$\mu((-\infty, x]) = \nu((-\infty, x]).$$

PROOF. The collection $\mathcal{C}$ of all intervals of the form $(-\infty, x]$ is a π-system generating $\mathcal{B}(\mathbb{R})$. Thus, the preceding proposition applies to prove sufficiency. Necessity is trivial. □

The following proposition extends Proposition 7.36 to σ-finite measures.

7.38 PROPOSITION. Let μ and ν be σ-finite measures on $(E, \mathcal{E})$. Suppose that they agree on a π-system $\mathcal{C}$ generating $\mathcal{E}$. Suppose further that there is a partition (E_n) of E such that $E_n \in \mathcal{C}$ and $\mu(E_n) = \nu(E_n) < \infty$ for every n. Then, $\mu = \nu$.

PROOF. For each n, define the measures μ_n and ν_n on $(E, \mathcal{E})$ by

$$\mu_n(A) = \mu(A \cap E_n), \quad \nu_n(A) = \nu(A \cap E_n), \quad A \in \mathcal{E}.$$

Since $E_n \in \mathcal{C}$, and since $A \cap E_n \in \mathcal{C}$ for every $A \in \mathcal{C}$, we have

$$\mu_n(A) = \mu(A \cap E_n) = \nu(A \cap E_n) = \nu_n(A) \text{ for } A \in \mathcal{C}.$$

And, by hypothesis, $\mu_n(E) = \mu(E_n) = \nu(E_n) = \nu_n(E) < \infty$. Thus, the last proposition applies to show that $\mu_n = \nu_n$ for each n. This completes the proof since $\mu = \sum \mu_n$ and $\nu = \sum \nu_n$.   □

Almost Everywhere

Often we face situations where a certain statement is true for every x in E_0 and E_0 is almost the same as E in the sense that $E_0 \in \mathcal{E}$ and $\mu(E \setminus E_0) = 0$. In that case, we say that the statement is true for *almost every* x in E or that the statement is true almost everywhere.

Incidentally, a set $N \subset E$ is said to be negligible if there is an A in $\mathcal{E}$ such that $N \subset A$ and $\mu(A) = 0$. So, a statement holds almost everywhere if and only if it fails only over a negligible set.

Examples

7.39 *Dirac measure.* Let $(E, \mathcal{E})$ be a measurable space. Fix $x \in E$. Define

$$\delta_x(A) = \begin{cases} 1 & \text{if } x \in A \\ 0 & \text{if } x \notin A \end{cases}$$

for each A in $\mathcal{E}$. Then, δ_x is a measure on $(E, \mathcal{E})$. It is called the *Dirac measure* sitting at x.

7.40 *Counting measures.* Let $(E, \mathcal{E})$ be a measurable space and let D be a countable subset of E. Define a measure ν on $(E, \mathcal{E})$ by

$$\nu = \sum_{x \in D} \delta_x.$$

Note that $\nu(A)$ is the number of points in $A \cap D$. Such measures are called counting measures.

7.41 Discrete measure spaces. Let E be countable and $\mathcal{E}$ be the collection of all subsets of E. Specifying a measure on $(E, \mathcal{E})$ is equivalent to assigning a number $m(x)$ in $\bar{\mathbb{R}}_+$ to each point x in E and then letting

$$\mu(A) = \sum_{x \in A} m(x), \quad A \in \mathcal{E}.$$

Then, m is called the mass function corresponding to μ. In particular, if $E = \{1, 2, \ldots, n\}$, every measure μ on $(E, \mathcal{E})$ can be regarded as a vector in $\mathbb{R}^n$.

7.42 Purely atomic measures. Let $(E, \mathcal{E})$ be a measurable space, let D be a countable subset of E, and let $m(x)$ be a positive number for each x in D. Define

$$\mu(A) = \sum_{x \in D} m(x) \delta_x(A), \quad A \in \mathcal{E}.$$

Then, μ is a measure on $(E, \mathcal{E})$. It puts the mass $m(x)$ at the point x, and there are only countably many such points x. Such measures μ are said to be purely atomic, and the points x with $\mu(\{x\}) > 0$ are called the atoms of μ provided that $\{x\} \in \mathcal{E}$.

Lebesgue Measures

These are the fundamental examples of measures that are familiar from calculus. A measure μ on $(\mathbb{R}, \mathcal{B}(\mathbb{R}))$ is called a *Lebesgue measure* on $\mathbb{R}$ if $\mu(A)$ is the length of A for every interval A; thus, μ is a generalization of the concept of length from intervals to Borel subsets of $\mathbb{R}$. Similarly, Lebesgue measure on $\mathbb{R}^2$ is the "area" measure, Lebesgue measure on $\mathbb{R}^3$ is the "volume" measure, and so on.

It is impossible to display $\mu(A)$ explicitly for each Borel set A, but countable additivity and various properties listed in Proposition 7.35 enable us to figure $\mu(A)$ out for most reasonable sets A.

More generally, given a Borel subset E of $\mathbb{R}^n$, it makes sense to talk of the Lebesgue measure on E; this is the restriction (see Exercise 7.47 below) of the Lebesgue measure on $\mathbb{R}^n$ to the trace space $(E, \mathcal{B}(E))$. These measures are unique. For instance, on $\mathbb{R}$, the collection $\mathcal{C}$ of all open intervals forms a π-system that satisfies the conditions of Proposition 7.38; thus, there can be at most one measure that assigns to each interval the length of that interval.

The existence of Lebesgue measure is another matter: countable additivity is to hold for every disjointed sequence of Borel sets. Since there are uncountably many such sequences, it is not clear whether such a measure exists. The following answers this question, albeit without proof.

7.43 THEOREM. There exists a unique measure on $(\mathbb{R}, \mathcal{B}(\mathbb{R}))$, called the Lebesgue measure, that assigns to each interval its length.

Let λ denote the Lebesgue measure on $(\mathbb{R}, \mathcal{B}(\mathbb{R}))$. The domain of λ can be extended slightly: let $\bar{\mathcal{B}}$ be the collection of all sets having the form $B \cup N$ where $B \in \mathcal{B}(\mathbb{R})$ and N is λ-negligible. Then, $\bar{\mathcal{B}}$ is again a σ-algebra on $\mathbb{R}$, and defining $\bar{\lambda} \colon \bar{\mathcal{B}} \mapsto \bar{\mathbb{R}}_+$ by putting $\bar{\lambda}(B \cup N) = \lambda(B)$ we obtain a measure $\bar{\lambda}$ on $(\mathbb{R}, \bar{\mathcal{B}})$. The elements of $\bar{\mathcal{B}}$ are called Lebesgue measurable sets, and $\bar{\lambda}$ is called the extended Lebesgue measure.

It seems impossible to extend λ much further. For instance, with $\mathcal{P}(E)$ denoting the collection of all subsets of E, we have the following, again without proof.

7.44 PROPOSITION. Let $E = [0, 1]$ and $\mathcal{E} = \mathcal{P}(E)$. Let μ be a finite measure on $(E, \mathcal{E})$. Then, there is a countable subset of D of E and positive numbers $m(x)$ for x in D such that

$$\mu(A) = \sum_{x \in D} m(x) 1_A(x), \qquad A \in \mathcal{E}.$$

As a corollary, we see that there can be no finite measure μ on $(E, \mathcal{E})$ such that $\mu(A)$ is the length of A for intervals A. Similarly, there can be no such Σ-finite measure either.

Exercises

7.45 *Infinite measures.* Let $(E, \mathcal{E})$ be as in Proposition 7.44. For A in $\mathcal{E}$, let $\mu(A)$ be the number of points in A; this is an integer if the set A is finite, and is $+\infty$ otherwise. Show that μ is a measure.

7.46 Show that $\mathcal{D}$ in the proof of Proposition 7.36 is a d-system.

7.47 *Restrictions.* Let $(E, \mathcal{E}, \mu)$ be a measure space. Let $D \in \mathcal{E}$ and let $\mathcal{D} = \{A \in \mathcal{E} : A \subset D\}$. Then, $(D, \mathcal{D})$ is the trace of $(E, \mathcal{E})$ on D. Define $\nu(A) = \mu(A)$ for A in $\mathcal{D}$. Then, ν is a measure on $(D, \mathcal{D})$; it is called the *restriction* of μ to D.

7.48 *Uniform distribution.* Let $D \subset \mathbb{R}$ be an interval of finite length. With λ Lebesgue measure on $\mathbb{R}$, let $\mu(B) = \lambda(B)/\lambda(D)$ for Borel subsets B of D. Show that μ is a probability measure on $(D, \mathcal{D})$ where $\mathcal{D} = \mathcal{B}(D)$. It is called the *uniform distribution* on D.

7.49 Σ-*Finiteness.* Let $E = \{a, b\}$ with the discrete σ-algebra, and define $\mu(\{a\}) = 0$, $\mu(\{b\}) = +\infty$. Show that this defines a Σ-finite measure μ that is not σ-finite.

7.50 *Atoms, atomic measures, diffuse measures.* Let $(E, \mathcal{E})$ be such that $\{x\} \in \mathcal{E}$ for every x in E. A point x is said to be an *atom* for the measure μ if $\mu(\{x\}) > 0$. If μ has no atoms, then it is said to be *diffuse*. If μ puts no mass outside the set of its atoms, then it is *purely atomic*. In general, μ will have some atomic part and some diffuse part. This is to show this decomposition.

    (a) Let μ be finite. Show that it has at most countably many atoms. Hint: Let D be the set of atoms, note that $D = \bigcup_n D_n$ where $D_n = \{x : \mu(\{x\}) \in [1/n,\ 1/(n-1))$, $n = 1, 2, \ldots$. Use the finiteness of μ to conclude that each D_n is a finite set, and therefore, that D must be countable.

    (b) Let μ be Σ-finite. Show that it has at most countably many atoms.

    (c) Let D be the set of atoms of a Σ-finite measure μ. Define

$$\nu(A) = \mu(A \cap D), \quad \lambda(A) = \mu(A \cap D^c), \quad A \in \mathcal{E}.$$

Then, ν is purely atomic, λ is diffuse, and

$$\mu = \nu + \lambda.$$

D. Integration

    Let $(E, \mathcal{E})$ be a measurable space. Recall that $\mathcal{E}$ stands also for the collection of all $\mathcal{E}$-measurable functions and that $\mathcal{E}_+$ is the subcollection consisting of positive $\mathcal{E}$-measurable functions. Given a measure μ on $(E, \mathcal{E})$, our aim is to define the "integral of f with respect to μ" for all reasonable functions f in $\mathcal{E}$. We shall denote it by any of the following:

$$\mu f = \int_E \mu(dx) f(x) = \int_E f \, d\mu.$$

When E is an interval of $\mathbb{R}$ and f is continuous and μ is the Lebesgue measure, the integral will coincide with the usual Riemann integral of f on E. When $E = \{1, \ldots, n\}$ and $\mathcal{E}$ is the discrete σ-algebra, every measure μ is specified by a row vector $(\mu_1, \ldots, \mu_n)$ with μ_i denoting $\mu(\{i\})$, and every function f in $\mathcal{E}$ corresponds to a column vector $(f_1, \ldots, f_n)$ with $f_i = f(i)$; in this case the integral μf will coincide with the product of the row vector $(\mu_1, \ldots, \mu_n)$ with the column vector with entries $f_1, \ldots, f_n$. As this last case illustrates, it is best to think of the integral as a product. After we define it, we shall show that it has the properties of products.

Definition of the Integral

    We define the integral μf in three steps: first for simple positive f, then for f in $\mathcal{E}_+$, finally for reasonable f in $\mathcal{E}$.

Step 1. Let f be a positive simple function. If its canonical form is $f = \sum_1^n a_i 1_{A_i}$, then we define

7.51
$$\mu f = \sum_1^n a_i \mu(A_i).$$

Step 2. Let $f \in \mathcal{E}_+$. Let (d_n) be defined by 7.18 and recall from the proof of Theorem 7.20 that $\lim d_n \circ f = f$. Now, for each n, the function $d_n \circ f$ is simple and positive, and the integral $\mu(d_n \circ f)$ is defined by the preceding step. We shall show in the remarks below that the numbers $\mu(d_n \circ f)$ form an increasing sequence, and hence, $\lim \mu(d_n \circ f)$ exists (it may be $+\infty$). Since $f = \lim d_n \circ f$, we define

7.52
$$\mu f = \lim \mu(d_n \circ f).$$

Step 3. Let $f \in \mathcal{E}$ arbitrary. Then, f^+ and f^- belong to $\mathcal{E}_+$, and their integrals are defined by the preceding step. Noting that $f = f^+ - f^-$, we define

7.53
$$\mu f = \mu f^+ - \mu f^-$$

provided that at least one term on the right is finite. Otherwise, if $\mu f^+ = \mu f^- = +\infty$, then μf does not exist.

7.54 REMARKS. (a) Formula 7.51 holds for positive simple functions even when $\sum_1^n a_i 1_{A_i}$ is not the canonical representation for f:

$$f = \sum_1^n a_i 1_{A_i} = \sum_1^m b_j 1_{B_j} \quad \Rightarrow \quad \mu f = \sum_1^n a_i \mu(A_i) = \sum_1^m b_j \mu(B_j).$$

This is easy to check using the finite additivity of μ.

(b) If f and g are positive simple functions and $a, b \in \mathbb{R}_+$, then $af + bg$ is again a positive simple function, and

$$\mu(af + bg) = a\,\mu f + b\,\mu g.$$

This can be checked using the preceding remark.

(c) If f is a positive simple function, then 7.51 shows that $\mu f \geq 0$ (it can be $+\infty$).

(d) If f and g are positive simple functions and $f \leq g$, then the preceding two remarks applied to f and $g - f$ show that $\mu f \leq \mu g$.

(e) In Step 2 of the definition, we have $d_1 \circ f \leq d_2 \circ f \leq \cdots$ and the preceding remark shows that $\mu(d_1 \circ f) \leq \mu(d_2 \circ f) \leq \cdots$ as claimed.

Integral over a Set

Let f be a measurable function, and A a measurable set. Then, $f1_A \in \mathcal{E}$. The *integral of f over A* is defined to be the integral of $f1_A$; it exists if and only if $\mu(f1_A)$ exists. The following notations are used for it:

7.55
$$\mu(f1_A) = \int_A \mu(dx)f(x) = \int_A f\, d\mu.$$

Integrability

A function f in $\mathcal{E}$ is said to be *integrable* if μf exists and is a real (i.e., finite) number. Thus, f is integrable if and only if $\mu f^+ < \infty$ and $\mu f^- < \infty$, or equivalently, if and only if $\mu|f| < \infty$ (note that $|f| = f^+ + f^-$). It is easy to see that, if f is integrable, then it is real-valued almost everywhere.

Elementary Properties

Here are some familiar properties of the integrals. A few others are put into the exercises.

7.56 PROPOSITION. (a) *Positivity.* If $f \in \mathcal{E}_+$, then $\mu f \geq 0$.

(b) *Monotonicity.* If $f, g \in \mathcal{E}_+$ and $f \leq g$, then $\mu f \leq \mu g$. If $f, g \in \mathcal{E}$ and f, g are integrable, and $f \leq g$, then $\mu f \leq \mu g$.

(c) *Finite additivity over sets.* Let $f \in \mathcal{E}_+$. If $\{A_1, \ldots, A_m\}$ is a measurable partition of A in $\mathcal{E}$, then

$$\int_A f\, d\mu = \sum_{i=1}^{m} \int_{A_i} f\, d\mu.$$

PROOF. (a) If $f \geq 0$, then the definition of μf yields $\mu f \geq 0$.

(b) If $0 \leq f \leq g$, then $d_n \circ f \leq d_n \circ g$ and so

$$\mu(d_n \circ f) \leq \mu(d_n \circ g)$$

by the monotonicity of integration for simple functions. Now, the left-hand side converges to μf and the right-hand side converges to μg. Hence $\mu f \leq \mu g$. The general case follows from the observation that if $f \leq g$ then $f^+ \leq g^+$ and $-f^- \leq -g^-$.

(c) Fix f in $\mathcal{E}_+$. Let $A_1, \ldots, A_m$ in $\mathcal{E}$ be disjoint with union A. If f is simple, the claim of (c) is immediate from Remark 7.54(b) applied to the simple functions

$f1_{A_1}, \ldots, f1_{A_m}$ whose sum is $f1_A$. Applying this to the simple functions $d_n \circ f$, we see that

$$\sum_1^m \mu(1_{A_i} d_n \circ f) = \mu(1_A d_n \circ f).$$

Note that $1_B(x) d_n \circ f(x) = d_n(1_B(x) f(x))$ for each x by the way the function d_n is defined. Putting this observation into the preceding expression and letting $n \to \infty$ we obtain

$$\begin{aligned}
\sum_1^m \mu(f1_{A_i}) &= \sum_1^m \lim_n \mu(d_n \circ (f1_{A_i})) \\
&= \lim_n \sum_1^m \mu(d_n \circ (f1_{A_i})) \\
&= \lim_n \mu(d_n \circ (f1_A)) = \mu(f1_A),
\end{aligned}$$

where the interchange of the limit and the sum is justified by the finiteness of m.   □

Of course, if $f = 0$ then $\mu f = 0$. The following converse is useful when μf is known but f is not.

7.57 COROLLARY. Let $f \in \mathcal{E}_+$. If $\mu f = 0$ then $f = 0$ almost everywhere.

PROOF. Let $f \in \mathcal{E}_+$ and $\mu f = 0$. We are to show that $\mu(A) = 0$ for $A = \{x : f(x) > 0\}$. Note that A is the limit of the increasing sequence of measurable sets $A_n = \{x : f(x) > 1/n\}$. Thus, $\mu(A) = \lim \mu(A_n)$ by the sequential continuity of measures; and $\mu(A_n) = 0$ for every n, since $0 \le \mu(A_n) = \mu 1_{A_n} \le n \mu f = 0$ by the monotonicity of integration and the observation that $1_{A_n} \le nf$.   □

Monotone Convergence Theorem

This is the key result in the theory of integration. It allows interchanging the order of taking limits and integrals under reasonable conditions.

7.58 THEOREM. Let $(f_n) \subset \mathcal{E}_+$ be increasing. Then,

$$\mu(\lim f_n) = \lim \mu f_n.$$

PROOF. Let $f = \lim f_n$; it is well-defined since $f_1 \le f_2 \le \cdots$, and is positive and $\mathcal{E}$-measurable. So, μf is well-defined. By the monotonicity of integration, $\mu f_1 \le \mu f_2 \le \cdots \le \mu f$. Therefore $\lim \mu f_n$ exists and

$$\lim_n \mu f_n \le \mu f.$$

It remains to show that $\lim_n \mu f_n \ge \mu f$. This is accomplished in steps.

Step 1. If $b \in \mathbb{R}_+$, $B \in \mathcal{E}$, and $f(x) > b$ for x in B, then $\lim_n \mu(f_n 1_B) \ge b \, \mu(B)$.
First, note that $\{f_1 > b\} \subset \{f_2 > b\} \subset \cdots$ and that

$$\bigcup_n \{f_n > b\} = \{x : f_n(x) > b \text{ for some } n\} = \{f > b\}.$$

Put $B_n = \{f_n > b\} \cap B$. Then, $B_n \nearrow \bigcup_n B_n = \{f > b\} \cap B = B$. Thus,

7.59 $$\lim_n \mu(B_n) = \mu(B)$$

by the sequential continuity of μ under increasing limits. Now, note that

$$f_n 1_B \ge f_n 1_{B_n} \ge b 1_{B_n},$$

and so the monotonicity of integration yields that

$$\mu(f_n 1_B) \ge \mu(b 1_{B_n}) = b \, \mu(B_n).$$

Taking limits on both sides and using 7.59, we get

7.60 $$\lim \mu(f_n 1_B) \ge b \, \mu(B).$$

Step 2. The same inequality holds even if $f(x) \ge b$ for every x in B.
For $b = 0$, this is trivial. For $b > 0$, apply Step 1 with $b - \epsilon$ to see that $\lim_n \mu(f_n 1_B) \ge (b - \epsilon) \, \mu(B)$. Since ϵ is arbitrary, we can let it go to zero to obtain the desired inequality.

Step 3. If g is a simple function and $g \le f$, then $\lim_n \mu f_n \ge \mu g$.
Let $\sum_1^m b_i 1_{B_i}$ denote the canonical representation for g. Then, our assumptions imply that $f(x) \ge g(x) = b_i$ for x in B_i. Hence, we may apply the result of Step 2 to conclude that

$$\lim_n \mu(f_n 1_{B_i}) \ge b_i \, \mu(B_i), \quad i = 1, \dots, m.$$

Hence, by Proposition 7.56(c) applied to the function f_n, we see that

$$
\begin{aligned}
\lim_n \mu f_n &= \lim_n \sum_1^m \mu(f_n 1_{B_i}) \\
&= \sum_1^m \lim \mu(f_n 1_{B_i}) \ge \sum_1^m b_i \, \mu(B_i) = \mu g.
\end{aligned}
$$

Step 4. $\lim_n \mu f_n \geq \mu f$.

Put $g = d_m \circ f$. Step 3 applied with this g yields $\lim_n \mu f_n \geq \mu(d_m \circ f)$. Letting $m \to \infty$ we get the desired result. $\square$

A particular consequence of the monotone convergence theorem is that, in the definition 7.52, the special sequence $(d_n \circ f)$ can be replaced by any sequence $(f_n) \subset \mathcal{E}_+$ increasing to f.

Linearity of Integration

7.61 PROPOSITION. If $f, g \in \mathcal{E}_+$ and $a, b \in \mathbb{R}_+$, then

$$\mu(af + bg) = a\,\mu f + b\,\mu g.$$

The same holds for arbitrary f, g in $\mathcal{E}$ and a, b in $\mathbb{R}$ provided that both sides are well-defined. It holds, in particular, if f and g are integrable.

PROOF. If f, g are simple, the result is established by direct checking as was remarked in Remark 7.54(b). For f, g in $\mathcal{E}_+$, and a, b in $\mathbb{R}_+$, choose (f_n) and (g_n) to be sequences of simple positive functions increasing to f and g, respectively. Then,

$$\mu(af_n + bg_n) = a\,\mu f_n + b\,\mu g_n,$$

and $af_n + bg_n \nearrow af + bg$, $f_n \nearrow f$, $g_n \nearrow f$. Taking limits on both sides and using the monotone convergence theorem completes the proof. If f, g in $\mathcal{E}$ are arbitrary, write $f = f^+ - f^-$ and $g = g^+ - g^-$ and go through the same steps. $\square$

Fatou's Lemma
This gives a useful inequality for arbitrary sequences of positive measurable functions.

7.62 LEMMA. Let $(f_n) \subset \mathcal{E}_+$. Then, $\mu(\liminf f_n) \leq \liminf \mu f_n$.

PROOF. Define $g_m = \inf_{n \geq m} f_n$. Then, $\liminf f_n$ is the limit of the increasing sequence $(g_m) \subset \mathcal{E}_+$, and thus

$$\mu(\liminf f_n) = \mu(\lim g_m) = \lim \mu g_m$$

by the monotone convergence theorem. On the other hand, $g_m \leq f_n$ for all $n \geq m$, which yields $\mu g_m \leq \mu f_n$ for all $n \geq m$, which in turn means that $\mu g_m \leq \inf_{n \geq m} \mu f_n$.

Hence, as needed,

$$\lim \mu g_m \leq \lim_{m} \inf_{n \geq m} \mu f_n = \liminf \mu f_n.$$

□

7.63 COROLLARY. (a) Let $(f_n) \subset \mathcal{E}$. If $f_n \geq g$ for all n for some integrable function g, then

$$\mu(\liminf f_n) \leq \liminf \mu f_n.$$

(b) Let $(f_n) \subset \mathcal{E}$. If $f_n \leq g$ for all n for some integrable function g, then

$$\mu(\limsup f_n) \geq \limsup \mu f_n.$$

PROOF. Suppose that the integrable function g is real-valued, so that $f_n - g$ and $g - f_n$ are well-defined. If $f_n \geq g$ for all n, then Fatou's lemma applies to the sequence of positive functions $f_n - g$ and the result is the claimed inequality. If $f_n \leq g$ for all n, then Fatou's lemma applies to $g - f_n$ and the desired inequality is obtained by noting that $\limsup r_n = -\liminf(-r_n)$ for every sequence (r_n) in $\bar{\mathbb{R}}$. If g is integrable, it is real-valued almost everywhere, that is, $\mu(E \setminus D) = 0$ for $D = \{x \in E : g(x) \in \mathbb{R}\}$, and the integrals of f_n and g on E are the same as their integrals over D; thus, the preceding proof applies to $f_n 1_D$ and $g 1_D$ and yields the desired conclusions. □

Dominated Convergence Theorem

This is the second important tool for interchanging the order of taking limits and integrals.

A function f is said to be dominated by a function g if $|f| \leq g$; note that $g \geq 0$ necessarily. A sequence of functions (f_n) is said to be *dominated* by g if $|f_n| \leq g$ for each n. If g can be taken to be a finite constant, then (f_n) is said to be *bounded*.

7.64 THEOREM. Suppose that $(f_n) \subset \mathcal{E}$ and is dominated by an integrable function g. If $\lim f_n$ exists, then it is integrable and

$$\mu(\lim_{n} f_n) = \lim_{n} \mu f_n.$$

PROOF. By assumption, $-g \leq f_n \leq g$ for every n, and g and $-g$ are both integrable. Thus, μf_n exists and is sandwiched between the finite numbers $-\mu g$ and μg. Now, both statements of the last corollary apply and we get

$$\mu(\liminf f_n) \leq \liminf \mu f_n \leq \limsup \mu f_n \leq \mu(\limsup f_n).$$

If $\lim f_n$ exists, then $\liminf f_n = \limsup f_n = \lim f_n$, and $\lim f_n$ is integrable since it is dominated by g. Hence, the extreme members of the preceding series of expressions are finite and equal, which means that equality holds throughout.   □

If $(f_n) \subset \mathcal{E}$ and is bounded, say by the constant b, and if the measure μ is finite, then we can take $g = b$ in the preceding theorem. The resulting corollary is called the *bounded convergence theorem*.

7.65 THEOREM. Let $(f_n) \subset \mathcal{E}$ and bounded. Suppose that μ is finite. If $\lim f_n$ exists, then

$$\mu(\lim_n f_n) = \lim_n \mu f_n.$$

7.66 REMARK. In Theorem 7.64, the condition that the sequence be dominated by some integrable function is necessary. Here is an example of what can happen otherwise. Let $(E, \mathcal{E}) = (\mathbb{R}_+, \mathcal{B}(\mathbb{R}_+))$ and let f_n be the sequence of functions shown in Fig. 7.1. Note that the sequence is not monotone and there is no integrable function that dominates them. Also, $\mu f_n = 1$ for all n and so $\lim \mu f_n = 1$, whereas $\lim f_n = 0$ and so $\mu \lim f_n = 0$.

Characterization of the Integral

The definition of the integral through 7.51–7.52 yields the integral μf for every f in $\mathcal{E}_+$. Thus, in effect, integration extends the domain of μ from the measurable sets to the space $\mathcal{E}_+$ of all positive $\mathcal{E}$-measurable functions (and beyond). Hence, we may regard μ as the mapping $f \mapsto \mu f$ from $\mathcal{E}_+$ into $\bar{\mathbb{R}}_+$. This mapping is positive, linear, and continuous under increasing limits; see Propositions 7.56(a), 7.58, and 7.61. Here is a summary and a useful converse.

7.67 THEOREM. Let $(E, \mathcal{E})$ be a measurable space. Let L be a mapping from $\mathcal{E}_+$ into $\bar{\mathbb{R}}_+$. Then, there exists a unique measure μ on $(E, \mathcal{E})$ such that $L(f) = \mu f$ for every f in $\mathcal{E}_+$ if and only if

(a) $f, g \in \mathcal{E}_+$ and $a, b \in \mathbb{R}_+ \implies L(af + bg) = aL(f) + bL(g)$,
(b) $(f_n) \subset \mathcal{E}_+$ and $f_n \nearrow f \implies L(f_n) \nearrow L(f)$.

PROOF. Necessity of the conditions follows from the properties of the integral μf. To show the sufficiency, suppose that the properties (a) and (b) hold for L. Define

$$\mu(A) = L(1_A), \qquad A \in \mathcal{E}.$$

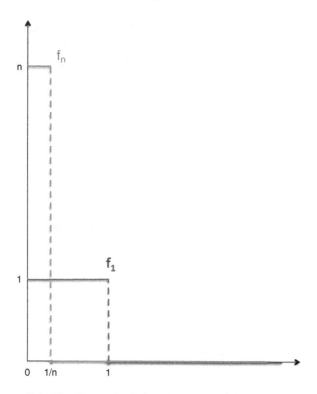

FIGURE 7.1. The first and nth functions in a sequence for which the dominated convergence theorem does not apply.

We show that μ is a measure. For the empty set $\emptyset$, we have $\mu(\emptyset) = 0$ because $L(f) = 0$ when $f = 0$ (as can be seen by taking $a = b = 0$ in the linearity property (a)). Second, if $A_1, A_2, \ldots$ are disjoint sets in $\mathcal{E}$ with union A, then the indicator of $\bigcup_1^n A_i$ is equal to $\sum_1^n 1_{A_i}$, the latter is increasing to 1_A, and hence

$$
\begin{aligned}
\mu(A) \;=\; L(1_A) \;&=\; \lim L\left(\sum_1^n 1_{A_i}\right) \\
&=\; \lim \sum_1^n L(1_{A_i}) \\
&=\; \lim \sum_1^n \mu(A_i) \;=\; \sum_1^\infty \mu(A_i),
\end{aligned}
$$

where we used the condition (b) to justify the second equality, and (a) to justify the third.

So, μ is a measure on $(E, \mathcal{E})$. It is unique by the necessity that $\mu(A) = L(1_A)$. Now, $L(f) = \mu f$ for simple f in $\mathcal{E}_+$ by the linearity property (a) for L and the linearity of integration. For arbitrary f in $\mathcal{E}_+$, choosing simple functions $f_n \nearrow f$,

$$L(f) = \lim L(f_n) = \lim \mu f_n = \mu f$$

by the condition (b) for L and the monotone convergence theorem for μ.   □

Lebesgue Versus Riemann

Let $E = [a, b]$, an interval on the real line. Let $\mathcal{E} = \mathcal{B}(E)$ and let λ be the Lebesgue measure on $(E, \mathcal{E})$. This is to discuss the Lebesgue integral

$$\lambda f = \int_{[a,b]} \lambda(dx) f(x)$$

versus the Riemann integral

$$Rf = \int_a^b f(x)\, dx$$

familiar from calculus. We limit the discussion to f in $\mathcal{E}_+$, that is, to positive Borel functions f.

Heuristically, the integral of f on E is the area under f over the interval E. If f is a step function, this has a clear meaning and is the same as λf and Rf. If f is continuous, it can be approximated by step functions, and, again, both λf and Rf exist and $\lambda f = Rf$. Similarly, if f is piecewise continuous. On the other hand, there are functions f for which λf exists but Rf does not: for example, suppose that $f(x)$ is equal to 1 or 0 accordingly as x is rational or irrational, that is, f is the indicator of the set $\mathbb{Q} \cap E$ of rational numbers in E; then, $\lambda f = \lambda(\mathbb{Q} \cap E) = 0$ since $\mathbb{Q}$ has Lebesgue measure 0, but the Riemann integral Rf does not exist. As this example illustrates, the Lebesgue integral exists for a larger class of functions than the Riemann integral does, and λf corresponds better to our intuitive expectations. The following is the complete picture; we omit the proof.

7.68 THEOREM. Let $f: [a, b] \mapsto \mathbb{R}$ and let D be the set of points of discontinuity for f. The Riemann integral Rf exists if and only if $\lambda(D) = 0$. If it exists, then so does the Lebesgue integral and the two integrals are the same.

Exercises

7.69　*Discrete spaces.* Let E, $\mathcal{E}$, μ be as in Example 7.41. Then, every $f\colon E \mapsto \mathbb{R}$ is $\mathcal{E}$-measurable. Show that, for $f\colon E \mapsto \mathbb{R}_+$,

$$\mu f = \sum_{x \in E} m(x) f(x).$$

7.70　*Purely atomic measures.* Let E, $\mathcal{E}$, D, μ be as in Example 7.42 and suppose that $\{x\} \in \mathcal{E}$ for every x in E (this is true for $E = \mathbb{R}^d$ and $\mathcal{E} = \mathcal{B}(\mathbb{R}^d)$ in particular). Show that, for f in $\mathcal{E}_+$,

$$\mu f = \sum_{x \in D} m(x) f(x).$$

7.71　*Integrability.* If $f \in \mathcal{E}_+$ and $\mu f < \infty$, then f is real-valued almost everywhere. Show this. More generally, if f is integrable, then f is real-valued almost everywhere.

7.72　*Monotone convergence.* If $(f_n) \subset \mathcal{E}_+$ then show that

$$\mu \sum_1^\infty f_n = \sum_1^\infty \mu f_n.$$

7.73　*Absolute values.* If f is integrable, then $|\mu f| \leq \mu |f|$. Show.

7.74　*Mean value theorem.* If $\mu(A) > 0$ and $a \leq f(x) \leq b$ for every x in A, then show that

$$a \leq \frac{1}{\mu(A)} \int_A f \, d\mu \leq b.$$

E.　Transforms and Indefinite Integrals

There are two basic methods of creating new measures from an old one. This section is to introduce them and their uses.

Image Measures

Let $(E, \mathcal{E})$ and $(F, \mathcal{F})$ be measurable spaces. Let μ be a measure on $(E, \mathcal{E})$, and let h be a transformation from E into F that is measurable with respect to $\mathcal{E}$ and $\mathcal{F}$.

Then, the inverse image $h^{-1}B$ belongs to $\mathcal{E}$ for every B in $\mathcal{F}$, and

7.75 $$\nu(B) = \mu(h^{-1}B), \qquad B \in \mathcal{F},$$

defines a measure ν on $(F, \mathcal{F})$. The new measure ν is called the *image* of μ under the transformation h; various notations used for ν are $\mu \circ h^{-1}$, μ_h, $h(\mu)$. The next theorem relates the integrals with respect to ν to the integrals with respect to the old measure μ.

7.76 THEOREM. For every f in $\mathcal{F}_+$, we have $\nu f = \mu(f \circ h)$.

PROOF. In integral notation, the defining relation 7.75 becomes

$$\nu 1_B = \mu(1_B \circ h),$$

since the indicator of $h^{-1}B$ is equal to $1_B \circ h$. Using this, if f is simple, say $f = \sum_1^n b_i 1_{B_i}$ with $B_1, \ldots, B_n$ in $\mathcal{F}$,

$$\nu f = \sum_i b_i \nu 1_{B_i} = \sum_i b_i \mu(1_{B_i} \circ h) = \mu \left(\sum_i b_i 1_{B_i} \circ h \right) = \mu(f \circ h).$$

If $f \in \mathcal{F}_+$, there is a sequence (f_n) of simple positive functions increasing to f, and the preceding applies to give $\nu f_n = \mu(f_n \circ h)$ for every n. Thus, taking limits as $n \to \infty$, we obtain the claimed formula via the monotone convergence theorem applied to each side separately.   □

The preceding theorem is a generalization of the change of variable formula from calculus. In more explicit notation, the claim is that

7.77 $$\int_F \nu(dy) f(y) = \int_E \mu(dx) f(h(x)),$$

that is, if $h(x)$ is replaced with y then the measure element $\mu(dx)$ is to be replaced with $\nu(dy)$. In calculus, often, $E = F = \mathbb{R}^d$ for some dimension $d \geq 1$, and μ and ν are measures that are expressed in terms of the Lebesgue measure and the Jacobian of the transformation h. See Exercises 7.86–7.88 for images of the Lebesgue measure on $\mathbb{R}$.

Indefinite Integrals

Let $(E, \mathcal{E}, \mu)$ be a measure space. Let p be a positive $\mathcal{E}$-measurable function. Define

7.78 $$\nu(A) = \int_A \mu(dx) p(x), \qquad A \in \mathcal{E}.$$

Then ν is a measure on $(E, \mathcal{E})$; this follows from the monotone convergence theorem for sums. We call ν the *indefinite integral* of p with respect to μ.

7.79 THEOREM. We have $\nu f = \mu(pf)$ for every f in $\mathcal{E}_+$.

PROOF. If f is an indicator, the claim is the definition 7.78 of ν. If f is simple, say $f = \sum_1^n a_i 1_{A_i}$, then

$$\nu f = \sum_i a_i \nu(A_i) = \sum_i a_i \int_E \mu(dx)p(x)1_{A_i}(x) = \mu(pf).$$

For arbitrary f in $\mathcal{E}_+$, there is a sequence of positive simple functions f_n increasing to f, and, by the monotone convergence theorem,

$$\nu f = \lim \nu f_n = \lim \mu(pf_n) = \mu(\lim pf_n) = \mu(pf).$$

$\square$

Written in explicit notation, the preceding theorem is that

7.80
$$\int_E \nu(dx)f(x) = \int_E \mu(dx)p(x)f(x)$$

for every f in $\mathcal{E}_+$. Obviously, this holds for arbitrary $\mathcal{E}$-measurable functions f for which the integrals are well-defined. Informally, this amounts to replacing $\nu(dx)$ with $\mu(dx)p(x)$, which we may write as

7.81
$$\nu(dx) = \mu(dx)p(x), \qquad x \in E.$$

Heuristically, we think of $\mu(dx)$ as the amount of mass put by μ on the "infinitesimal neighborhood" dx of the point x, and similarly for $\nu(dx)$. Then, 7.81 expresses $p(x)$ as the mass density at x of the measure ν with respect to μ. Thus, the function p is called the *density function* of ν with respect to μ, and the following notations are used for it:

7.82
$$p = \frac{d\nu}{d\mu}; \qquad p(x) = \frac{\nu(dx)}{\mu(dx)}, \quad x \in E.$$

The expressions 7.80–7.82 are equivalent ways of saying the same thing.

Radon–Nikodym Theorem

Let μ and ν be measures on a measurable space $(E, \mathcal{E})$. Then, ν is said to be absolutely continuous with respect to μ if, for every set A in $\mathcal{E}$,

7.83
$$\mu(A) = 0 \Longrightarrow \nu(A) = 0.$$

If ν is an indefinite integral as in 7.78, then it is absolutely continuous with respect to μ. The following, called the *Radon–Nikodym theorem*, provides a converse. We state this here without proof.

7.84 THEOREM. Suppose that μ is σ-finite, and that ν is absolutely continuous with respect to μ. Then, there exists a positive $\mathcal{E}$-measurable function p such that

7.85 $$\int_E \nu(dx)f(x) = \int_E \mu(dx)p(x)f(x), \qquad f \in \mathcal{E}_+.$$

Moreover, p is unique up to almost sure equality, that is, if $\hat{p}$ is in $\mathcal{E}_+$ and 7.85 holds with $\hat{p}$, then $\hat{p}(x) = p(x)$ for μ-almost every x in E.

The function p in the preceding theorem can be denoted by $d\nu/d\mu$ in view of the equivalence of 7.80–7.82; and thus, p is called the Radon–Nikodym derivative of ν with respect to μ.

Exercises

7.86 *Distribution functions and quantiles.* Let μ be a measure on $(\mathbb{R}, \mathcal{B}(\mathbb{R}))$ such that $c(x) = \mu(-\infty, x]$ is finite for every x. Then, c is called the *cumulative distribution function* of the measure μ, especially when μ is a probability measure. Show that c is an increasing right-continuous function. Define

$$q(u) = \inf\{x \in \mathbb{R} : c(x) > u\}, \qquad u \in [0, b],$$

where $b = \lim_{x\to\infty} c(x) = \mu(\mathbb{R})$; it is possible that $b = +\infty$. Show that q is increasing and right-continuous; it is called the *quantile function* corresponding to c. When c is continuous and strictly increasing, $q(u) = x$ if and only if $c(x) = u$, that is, q and c are functional inverses of each other.

7.87 *Images of Lebesgue.* Let μ, c, q, b be as in the preceding exercise. Let λ be the Lebesgue measure on the interval $(0, b)$ with its Borel σ-algebra. Show that

$$\mu(A) = \lambda(q^{-1}A), \qquad A \in \mathcal{B}(\mathbb{R}).$$

Thus, most measures μ on $(\mathbb{R}, \mathcal{B}(\mathbb{R}))$ are images of Lebesgue measures on intervals. Of course, then, by Theorem 7.76,

$$\int_{\mathbb{R}} \mu(dx)f(x) = \int_0^b \lambda(du)f(q(u))$$

for every positive Borel function f on $\mathbb{R}$ (and positivity can be removed if f is integrable). This reduces integrals with respect to μ to integrals with respect to the

Lebesgue measure; and the latter integral is equal to a Riemann integral if f is continuous or piecewise continuous.

As a corollary, this exercise shows the following. If c is an increasing right-continuous function on $\mathbb{R}$, then there exists a measure μ on $(\mathbb{R}, \mathcal{B}(\mathbb{R}))$ whose distribution function is c.

7.88 *Continuation.* This is to extend the preceding to measures μ that are Σ-finite, say, $\mu = \sum_1^\infty \mu_n$ where each μ_n is a finite measure on $(\mathbb{R}, \mathcal{B}(\mathbb{R}))$. Let $b_n = \mu_n(\mathbb{R})$ and define c_n and q_n as in the preceding exercise but for μ_n. Let $a_0 = 0$, $a_n = \sum_1^n b_m$ for $n \geq 1$, and $q(u) = q_{n+1}(u - a_n)$ for u in $[a_n, a_{n+1})$. Show that, then, $\mu = \lambda \circ q^{-1}$ where λ is the Lebesgue measure on $(\mathbb{R}_+, \mathcal{B}(\mathbb{R}_+))$. Again, then, the integral μf becomes the Lebesgue integral of $f \circ q$ on $\mathbb{R}_+$.

7.89 *Radon–Nikodym derivatives.* Let μ be a measure on $(\mathbb{R}, \mathcal{B}(\mathbb{R}))$ such that $c(x) = \mu(-\infty, x]$ is a real number for every x in $\mathbb{R}$. Suppose that μ is absolutely continuous with respect to the Lebesgue measure λ. Then, the function c is differentiable at λ-almost every point x, and

$$p(x) = \frac{\mu(dx)}{\lambda(dx)} = \frac{d}{dx}c(x) \qquad \text{for } \lambda\text{-almost every } x.$$

7.90 *Singularity.* Let μ and ν be measures on some measurable space $(E, \mathcal{E})$. Then, ν is said to be *singular* with respect to μ if there exists a set D in $\mathcal{E}$ such that $\mu(D) = 0$ and $\nu(E \setminus D) = 0$. The notion is the opposite of absolute continuity. For example, if μ is diffuse and ν purely atomic, then ν is singular with respect to μ.

7.91 *Cantor set, Cantor measure.* Consider the Cantor set C of Example 2.40, and recall the set D that consists of open intervals. Show that $\lambda(D) = 1$ and $\lambda(C) = 0$, where λ is the Lebesgue measure on $[0, 1]$. Let g be as in Example 2.40, and define $\mu = \lambda \circ g^{-1}$, where λ is the Lebesgue measure on $[0, 1)$. Then, $\mu(C) = 1$ and $\mu(D) = 0$. The measure μ is called the Cantor measure; it is singular with respect to the Lebesgue measure, and its distribution function is the function f of Example 2.40.

7.92 *Lebesgue–Stieltjes integrals.* Let $g\colon \mathbb{R} \mapsto \mathbb{R}_+$ be increasing and right-continuous. Let μ be the measure on $\mathbb{R}$ whose cumulative distribution function is g; see Exercise 7.87. For positive Borel functions f on $\mathbb{R}$, define

$$\int_{\mathbb{R}} f(x)\, dg(x) = \int_{\mathbb{R}} \mu(dx) f(x).$$

The left side is called the *Lebesgue–Stieltjes* integral of f with respect to the function g; the right side is the integral μf. This can be extended to arbitrary Borel f as usual by using the decomposition $f = f^+ - f^-$, assuming that either μf^+ or μf^- is finite. Integration can be extended to functions g that can be decomposed as $g = g_1 - g_2$, where both g_1 and g_2 are increasing right-continuous:

$$\int_{\mathbb{R}} f(x)\, dg(x) = \int_{\mathbb{R}} f(x)\, dg_1(x) - \int_{\mathbb{R}} f(x)\, dg_2(x),$$

for those f for which the integrals on the right side both make sense and are not both $+\infty$ or both $-\infty$.

7.93 *Functions of bounded variation.* These are functions that can be written as the difference of two increasing functions. For $g \colon \mathbb{R} \mapsto \mathbb{R}$, we define

$$V_g(x,y) = \sup_{\mathcal{A}} \sum_i |g(x_{i+1}) - g(x_i)|,$$

where the supremum is taken over all subdivisions $\mathcal{A}$ of $[x,y]$, where $\mathcal{A}$ is a finite collection of intervals $(x_0, x_1], (x_1, x_2], \ldots, (x_{n-1}, x_n]$ with $x = x_0 < x_1 < x_2 < \cdots < x_n = y$. The number $V_g(x,y)$ is called the *total variation* of g over the interval $[x,y]$, and g is said to be of *bounded variation* over $[x,y]$ if $V_g(x,y) < \infty$. Show the following:

(a) If g is increasing, then $V_g(x,y) = g(y) - g(x)$.
(b) If $x < y < z$, then $V_g(x,y) + V_g(y,z) = V_g(x,z)$.
(c) If $g = g_1 + g_2$, then $V_g(x,y) \leq V_{g_1}(x,y) + V_{g_2}(x,y)$.
(d) The function g is of bounded variation over $[x,y]$ if and only if $g = g_1 - g_2$, where g_1 and g_2 are increasing. Hint: For u in $(x,y]$, define g_1 and g_2 such that

$$2g_1(u) = V_g(x,u) + g(u) + g(x), \qquad 2g_2(u) = V_g(x,u) - g(u) + g(x).$$

F. Kernels and Product Spaces

Let $(E, \mathcal{E})$ and $(F, \mathcal{F})$ be measurable spaces. A mapping $K \colon E \times \mathcal{F} \mapsto \bar{\mathbb{R}}_+$ is called a *transition kernel* from $(E, \mathcal{E})$ to $(F, \mathcal{F})$ if

(a) the mapping $x \mapsto K(x, B)$ is $\mathcal{E}$-measurable for each B in $\mathcal{F}$, and
(b) the mapping $B \mapsto K(x, B)$ is a measure on $(F, \mathcal{F})$ for each x in E.

It will be convenient to write K_x for the measure $B \mapsto K(x, B)$.

For example, with $E = F = \mathbb{R}$ and $\mathcal{E} = \mathcal{F} = \mathcal{B}(\mathbb{R})$, putting

$$K(x, B) = \int_B dy\, \frac{e^{-(y-x)^2/2}}{\sqrt{2\pi}}, \qquad x \in \mathbb{R},\ B \in \mathcal{B}(\mathbb{R}),$$

yields a transition kernel K, which is further a probability kernel. For another example, let $E = \{1, 2, \ldots, m\}$ and $F = \{1, 2, \ldots, n\}$ with their discrete σ-algebras; then, a transition kernel from $(E, \mathcal{E})$ to $(F, \mathcal{F})$ is specified by the numbers $K(x, \{y\})$, and we may regard K as an $m \times n$ matrix of positive numbers. Thus, transition kernels are generalizations of positive matrices; recall that functions are generalizations of column vectors and measures are generalizations of row vectors.

Functions on Product Spaces

If $f: E \times F \mapsto \mathbb{R}$, then the mapping $y \mapsto f(x, y)$ is called the *section* of f at the point x of E, and, similarly, $x \mapsto f(x, y)$ is called the section of f at the point y of F. The following shows that sections of a measurable function are measurable.

7.94 THEOREM. Let $f: E \times F \mapsto \mathbb{R}$ be $\mathcal{E} \otimes \mathcal{F}$-measurable. Then, $y \mapsto f(x, y)$ is $\mathcal{F}$-measurable for each x in E, and $x \mapsto f(x, y)$ is $\mathcal{E}$-measurable for each y in F.

PROOF. Fix x_0 in E and let $h: y \mapsto f(x_0, y)$; we are to show that h is $\mathcal{F}$-measurable. Observe that $h = f \circ g$, where $g: F \mapsto E \times F$ is defined by setting $g(y) = (x_0, y)$. For a measurable rectangle $A \times B$ in $\mathcal{E} \otimes \mathcal{F}$, the inverse image $g^{-1}(A \times B)$ is either $\emptyset$ or B, either case being in $\mathcal{F}$. Since the measurable rectangles generate $\mathcal{E} \otimes \mathcal{F}$, this shows that g is measurable with respect to $\mathcal{F}$ and $\mathcal{E} \otimes \mathcal{F}$. Hence, the composition $h = f \circ g$ of measurable functions f and g is $\mathcal{F}$-measurable. Since x_0 in E is arbitrary, this proves the measurability of $y \mapsto f(x, y)$ for every x. The measurability of $x \mapsto f(x, y)$ is proved by symmetry. □

Unfortunately, the converse to the preceding proposition is generally false: it is possible that $x \mapsto f(x, y)$ is $\mathcal{E}$-measurable for every y and $y \mapsto f(x, y)$ is $\mathcal{F}$-measurable for every x, and yet f fails to be $\mathcal{E} \otimes \mathcal{F}$-measurable. If $f(x, y) = g(x)h(y)$ for some $\mathcal{E}$-measurable function g and some $\mathcal{F}$-measurable function h, then f is $\mathcal{E} \otimes \mathcal{F}$-measurable. In general, one needs some such special property to conclude that f is $\mathcal{E} \otimes \mathcal{F}$-measurable; see Exercise 7.105 for an example.

Kernels–Functions

7.95 PROPOSITION. Let K be a transition kernel from $(E, \mathcal{E})$ to $(F, \mathcal{F})$. For positive $\mathcal{E} \otimes \mathcal{F}$-measurable functions f, define

$$Tf(x) = \int_F K(x, dy) f(x, y), \qquad x \in E.$$

Then, Tf is a positive $\mathcal{E}$-measurable function.

PROOF. Let f be such. For x in E, its section at x is the $\mathcal{F}$-measurable function $f_x: y \mapsto f(x,y)$. Note that $Tf(x)$ is the integral of f_x with respect to the measure K_x. Thus, Tf is well-defined; and, by the linearity of integration and the monotone convergence theorem,

7.96 $$T(af + bg) = a\,Tf + b\,Tg, \qquad \lim Tf_n = Tf,$$

the latter for (f_n) increasing to f. In view of these properties, proving that Tf is $\mathcal{E}$-measurable reduces to showing that Tf is $\mathcal{E}$-measurable for f that are indicators of the measurable rectangles $A \times B$; recall that the measurable rectangles form a π-system generating the product σ-algebra and the monotone class theorem applies to the class of f for which Tf is $\mathcal{E}$-measurable. But for A in $\mathcal{E}$ and B in $\mathcal{F}$,

$$T1_{A\times B}(x) = \int_F K(x,dy)1_A(x)1_B(y) = 1_A(x)K(x,B), \qquad x \in E,$$

which is $\mathcal{E}$-measurable since both 1_A and $x \mapsto K(x,B)$ are such. $\qquad\square$

In particular, if $f(x,y)$ is free of x, we prefer to write Kf instead of Tf, thereby giving full play to the matrix–column vector analogy; that is, for $f: F \mapsto \mathbb{R}$ that is positive $\mathcal{F}$-measurable,

$$Kf(x) = \int_F K(x,dy)f(y), \qquad x \in E.$$

Measures on the Product Space

The following is the general method for constructing measures on the product space $(E \times F, \mathcal{E} \otimes \mathcal{F})$.

7.97 THEOREM. Let μ be a measure on $(E, \mathcal{E})$. Let K be a transition kernel from $(E, \mathcal{E})$ to $(F, \mathcal{F})$. Then, there is a unique measure π on $(E \times F, \mathcal{E} \otimes \mathcal{F})$ such that

7.98 $$\int_{E\times F} \pi(dx\,dy)f(x,y) = \int_E \mu(dx) \int_F K(x,dy)f(x,y)$$

for every positive $\mathcal{E} \otimes \mathcal{F}$-measurable function f. In particular,

7.99 $$\pi(A \times B) = \int_A \mu(dx)K(x,B), \qquad A \in \mathcal{E}, B \in \mathcal{F}.$$

PROOF. For positive $\mathcal{E} \otimes \mathcal{F}$-measurable f, the right side of 7.98 is equal to $\mu(Tf)$, the integral of the function Tf of the last proposition with respect to the measure μ. The linearity of the integration with respect to μ and the monotone convergence theorem yield, together with 7.96, that

$$\mu(T(af + bg)) = \mu(a\,Tf + b\,Tg) = a\,\mu(Tf) + b\,\mu(Tg)$$

and

$$\lim \mu(Tf_n) = \mu(Tf) \qquad \text{if } f_n \nearrow f.$$

It follows from Theorem 7.67 that there is a unique measure π on $(E \times F, \mathcal{E} \otimes \mathcal{F})$ such that $\mu(Tf)$ is equal to the integral of f with respect to π for every positive f in $\mathcal{E} \otimes \mathcal{F}$. In particular, for $f = 1_{A \times B}$ with A in $\mathcal{E}$ and B in $\mathcal{F}$, we obtain the claimed formula for $\pi(A \times B)$.   □

In the preceding theorem, the measure π defined through its integral by 7.98 is unique, and it satisfies 7.99. But there might be other measures that satisfy 7.99 but fail to satisfy 7.98 for some f. The following is directed at this matter of uniqueness of π satisfying 7.99. We call K σ-*bounded* if there exists an $\mathcal{F}$-measurable partition (F_n) of F such that $x \mapsto K(x, F_n)$ is a bounded function for each n.

7.100 COROLLARY. In the preceding theorem, suppose that μ is σ-finite and K is σ-bounded. Then, there is exactly one measure π satisfying 7.99, and it is σ-finite and it satisfies 7.98.

PROOF. Let (E_m) be an $\mathcal{E}$-measurable partition of E such that $\mu(E_m)$ is finite for each m. Let (F_n) be an $\mathcal{F}$-measurable partition of F such that $x \mapsto K(x, F_n)$ is bounded for each n. Then, $\pi(E_m \times F_n)$ defined by 7.99 is finite for each pair (m, n), and the measurable rectangles $E_m \times F_n$ form a partition of $E \times F$.

Let $\hat{\pi}$ be a measure on the product space satisfying 7.99. Then, $\pi(A \times B) = \hat{\pi}(A \times B)$ for every measurable rectangle, and there is a partition $(E_m \times F_n)$ of $E \times F$ for which $\pi(E_m \times F_n) = \hat{\pi}(E_m \times F_n) < \infty$. It follows from Proposition 7.38 that $\hat{\pi} = \pi$.   □

Product Measures and Fubini

In the last theorem, the measure π is often denoted by $\mu \times K$. In particular, if the kernel K has the special form $K(x, B) = \nu(B)$ for some measure ν on $(F, \mathcal{F})$, then π is called the product of μ and ν and is denoted by $\mu \times \nu$. Note that, then, $\pi(A \times B) = \mu(A)\nu(B)$ for A in $\mathcal{E}$ and B in $\mathcal{F}$, and the condition of σ-boundedness on K is equivalent to the σ-finiteness of the measure ν.

The next theorem, called Fubini's, is concerned with integration with respect to the product measure $\pi = \mu \times \nu$. Its main point is the formula 7.102: under reasonable conditions, in repeated integration, one can change the order of integration without harm. Recall that $\mathcal{E} \otimes \mathcal{F}$ stands also for the collection of all $\mathcal{E} \otimes \mathcal{F}$-measurable functions.

7.101 THEOREM. Let μ and ν be σ-finite measures on $(E, \mathcal{E})$ and $(F, \mathcal{F})$, respectively. There exists a unique σ-finite measure π on $(E \times F, \mathcal{E} \otimes \mathcal{F})$ such that, for every positive f in $\mathcal{E} \otimes \mathcal{F}$,

7.102 $$\pi f = \int_E \mu(dx) \int_F \nu(dy) f(x, y) = \int_F \nu(dy) \int_E \mu(dx) f(x, y).$$

This holds for arbitrary f in $\mathcal{E} \otimes \mathcal{F}$ provided that f is π-integrable, that is, $\pi(f^+)$ and $\pi(f^-)$ are both finite; moreover, then, $y \mapsto f(x, y)$ is ν-integrable for μ-almost every x, and $x \mapsto f(x, y)$ is μ-integrable for ν-almost every y.

7.103 REMARK. Since we have more than one measure, we need to indicate the measure involved for notions like integrability and negligibility. So, μ-almost every x means for every x outside a set N in $\mathcal{E}$ with $\mu(N) = 0$, and ν-integrable means integrable with respect to the measure ν, etc.

PROOF. The measure $\pi = \mu \times \nu$; its existence, uniqueness, and σ-finiteness follow from Theorem 7.97 and Corollary 7.100 with $K(x, B) = \nu(B)$.

We show next that, for f positive and $\mathcal{E} \otimes \mathcal{F}$-measurable,

7.104 $$\int_E \mu(dx) \int_F \nu(dy) f(x, y) = \int_F \nu(dy) \int_E \mu(dx) f(x, y).$$

The left side is equal to πf where $\pi = \mu \times \nu$ as mentioned above. The right side is equal to $\hat{\pi}\hat{f}$, where $\hat{\pi} = \nu \times \mu$ on $(F \times E, \mathcal{F} \otimes \mathcal{E})$ and $\hat{f}(y, x) = f(x, y)$. Note that $f = \hat{f} \circ h$, where $h: E \times F \mapsto F \times E$ is the transposition mapping $(x, y) \mapsto (y, x)$. It is obvious that h is measurable with respect to $\mathcal{E} \otimes \mathcal{F}$ and $\mathcal{F} \otimes \mathcal{E}$. For A in $\mathcal{E}$ and B in $\mathcal{F}$,

$$\pi(h^{-1}(B \times A)) = \pi(A \times B) = \mu(A)\nu(B) = \hat{\pi}(B \times A).$$

Assuming that μ and ν are finite, this equality of $\hat{\pi}$ and $\pi \circ h^{-1}$ on the π-system of measurable rectangles implies that $\hat{\pi} = \pi \circ h^{-1}$. Thus,

$$\hat{\pi}\hat{f} = (\pi \circ h^{-1})\hat{f} = \pi(\hat{f} \circ h) = \pi f.$$

This establishes 7.104 for μ and ν finite and f positive.

Assume, next, that μ and ν are σ-finite. Let (E_m) be a measurable partition of E such that $\mu(E_m) < \infty$ for every m; similarly, pick (F_n) such that $\nu(F_n) < \infty$ for every n. Then, the preceding step yields

$$\int_{E_m} \mu(dx) \int_{F_n} \nu(dy) f(x, y) = \int_{F_n} \nu(dy) \int_{E_m} \mu(dx) f(x, y)$$

for every m and n. Summing both sides over all m and n gives the result 7.104 claimed for f positive.

Dropping the condition of positivity on f, suppose next that f is π-integrable. Then, 7.104 holds for f^+ and f^- separately, and $\pi f = \pi(f^+) - \pi(f^-)$ with both terms finite. Hence 7.104 holds.

As to the integrability of sections, observe that the integrability of f implies that $x \mapsto \int \nu(dy) f(x, y)$ is real-valued for μ-almost every x, which in turn implies that $y \mapsto f(x, y)$ is ν-integrable for μ-almost every x. By symmetry, the finiteness of the right side of 7.104 implies that $x \mapsto f(x, y)$ is μ-integrable for ν-almost every y.   □

Infinite Product Spaces

The concepts and results can be extended by induction to products of finitely many spaces: let $(E_1, \mathcal{E}_1), \ldots, (E_n, \mathcal{E}_n)$ be measurable spaces. Their product is denoted by

$$\bigotimes_{i=1}^{n} (E_i, \mathcal{E}_i) = (E_1 \times \cdots \times E_n, \mathcal{E}_1 \otimes \cdots \otimes \mathcal{E}_n),$$

where $E_1 \times \cdots \times E_n$ is the set of all n-tuples $(x_1, \ldots, x_n)$ with x_i in E_i for each i, and $\mathcal{E}_1 \otimes \cdots \otimes \mathcal{E}_n$ is the σ-algebra generated by the *measurable rectangles* $A_1 \times \cdots \times A_n$ with A_i in $\mathcal{E}_i$ for each i.

Measures on the product space are obtained by repeated applications of Theorem 7.97. We illustrate the technique with $n = 3$. Let μ_1 be a measure on $(E_1, \mathcal{E}_1)$, let K_2 be a transition kernel from $(E_1, \mathcal{E}_1)$ to $(E_2, \mathcal{E}_2)$, and let K_3 be a transition kernel from $(E_1 \times E_2, \mathcal{E}_1 \otimes \mathcal{E}_2)$ to $(E_3, \mathcal{E}_3)$. Then, the formula

$$\pi f = \int_{E_1} \mu_1(dx_1) \int_{E_2} K_2(x_1, dx_2) \int_{E_3} K_3((x_1, x_2), dx_3) f(x_1, x_2, x_3)$$

for positive f in $\mathcal{E}_1 \otimes \mathcal{E}_2 \otimes \mathcal{E}_3$ defines a measure on $(E_1 \times E_2 \times E_3, \mathcal{E}_1 \otimes \mathcal{E}_2 \otimes \mathcal{E}_3)$. In short, $\pi = (\mu_1 \times K_2) \times K_3$, which can be written as

$$\pi = \mu \times K_2 \times K_3.$$

New issues arise when the product space is that of infinitely many measurable spaces, especially when there are uncountably many spaces. Let T be an arbitrary set, countable or uncountable; it will play the role of an index set, we think of it as the time set (this of the special cases where $T = \mathbb{N}$ or $T = \mathbb{R}_+$). For each t in T, let $(E_t, \mathcal{E}_t)$ be a measurable space. Let x_t be a point in E_t for each t. We write $(x_t)_{t \in T}$ for the resulting collection and think of it as a function on T; this is especially appropriate when $(E_t, \mathcal{E}_t) = (E, \mathcal{E})$ for all t, in which case $x = (x_t)_{t \in T}$ can be regarded as a function $t \mapsto x_t$ from T into E. The set F of all such functions $x = (x_t)_{t \in T}$ is called the *product space* defined by $\{E_t : t \in T\}$. On the set F, we introduce the product σ-algebra $\mathcal{F}$ generated by the collection of all measurable rectangles

$$\underset{t \in T}{\times} A_t = \{x \in F : x_t \in A_t \text{ for each } t \text{ in } T\},$$

where A_t differs from E_t for only finitely many t, and $A_t \in \mathcal{E}_t$ for those t. The resulting measurable space $(F, \mathcal{F})$ is denoted as

$$(F, \mathcal{F}) = \bigotimes_{t \in T}(E_t, \mathcal{E}_t),$$

and it is usual to write $\bigotimes_{t \in T} \mathcal{E}_t$ for $\mathcal{F}$. In the special case where $(E_t, \mathcal{E}_t) = (E, \mathcal{E})$ for all t, one writes $(E, \mathcal{E})^{\mathrm{T}}$ or $(E^{\mathrm{T}}, \mathcal{E}^{\mathrm{T}})$ for $(F, \mathcal{F})$; and when $T = \{1, 2, \dots\}$, it is usual to write $(E, \mathcal{E})^\infty = (E^\infty, \mathcal{E}^\infty)$ for the same.

Such product spaces occur naturally in probability theory. There, each x in F is a possible path for some particle moving through the spaces E_t over time. The probability measure π on $(F, \mathcal{F})$ then assigns to the set B in $\mathcal{F}$ the chances that the particle's path is a function belonging to B. Often, π is to be constructed from "finite-dimensional" data, the latter giving $\pi(B)$ for B that are measurable rectangles. The details of such a construction are outside the scope of this book.

Exercises

7.105 *Measurability.* Suppose that $E = \mathbb{R}$, $\mathcal{E} = \mathcal{B}(\mathbb{R})$, and $(F, \mathcal{F})$ is arbitrary. Let $f \colon E \times F \mapsto \mathbb{R}$. Suppose that $x \mapsto f(x, y)$ is right-continuous for each y, and $y \mapsto f(x, y)$ is $\mathcal{F}$-measurable for each x. Then, show that f is $\mathcal{E} \otimes \mathcal{F}$-measurable.

7.106 *Images and kernels.* Let $(E, \mathcal{E})$ and $(F, \mathcal{F})$ be measurable spaces, and let $h \colon E \mapsto F$ be measurable with respect to $\mathcal{E}$ and $\mathcal{F}$. Define

$$K(x, B) = 1_B \circ h(x), \qquad x \in E, \ B \in \mathcal{F}.$$

Show that K is a transition probability kernel. For f in $\mathcal{F}_+$, show that $Kf(x) = f \circ h(x)$.

7.107 *Transition densities.* Let ν be a σ-finite measure on $(F, \mathcal{F})$, and let k be a positive function in $\mathcal{E} \otimes \mathcal{F}$. Define K by

$$K(x, B) = \int_B \nu(dy) k(x, y), \qquad x \in E, \ B \in \mathcal{F}.$$

Show that K is a transition kernel. Then, k is called the transition density of K with respect to ν.

7.108 *Measure–kernel–function.* Let K be a transition kernel from $(E, \mathcal{E})$ to $(F, \mathcal{F})$. For f in $\mathcal{F}_+$ define

$$Kf(x) = \int_F K(x, dy) f(y), \qquad x \in E,$$

and show that $Kf \in \mathcal{E}_+$. For a measure μ on $(E, \mathcal{E})$, let

$$\mu K(B) = \int_E \mu(dx) K(x, B), \qquad B \in \mathcal{F},$$

and show that μK is a measure on $(F, \mathcal{F})$. Show that $(\mu K)f = \mu(Kf)$. Interpret these when E and F are finite spaces.

7.109 *Products of kernels.* Let $(E, \mathcal{E})$, $(F, \mathcal{F})$, and $(G, \mathcal{G})$ be measurable spaces. Let K be a transition kernel from $(E, \mathcal{E})$ to $(F, \mathcal{F})$, and L a transition kernel from $(F, \mathcal{F})$ to $(G, \mathcal{G})$. Define

$$M(x, B) = \int_F K(x, dy) L(y, B), \qquad x \in E, \ B \in \mathcal{G}.$$

Show that M is a transition kernel from $(E, \mathcal{E})$ to $(G, \mathcal{G})$. It is usual to write KL for M. When $(E, \mathcal{E}) = (F, \mathcal{F}) = (G, \mathcal{G})$, and $K = L$, then one writes K^2 for KL.

Further Reading

The treatment here follows Rudin [Rud76] and Royden [Roy88] closely, especially on the classic material about metric spaces. On measure theory, we followed Halmos [Hal74] and Cohn [Coh94]. Convex analysis material is based on Rockafellar [Roc70]. The following bibliography lists the reference books and related literature.

E. Çınlar and R.J. Vanderbei, *Real and Convex Analysis*, Undergraduate Texts in Mathematics, DOI 10.1007/978-1-4614-5257-7, © Springer Science+Business Media New York 2013

Bibliography

[BL06]   J.M. Borwein and A.S. Lewis. *Convex Analysis and Nonlinear Optimization: Theory and Examples*. Springer, New York, 2nd edition, 2006.

[Coh94] D.L. Cohn. *Measure Theory*. Birkhäuser, Boston, 1994.

[Hal74] P.R. Halmos. *Measure Theory*. Springer, New York, 1974.

[Roc70] R.T. Rockafellar. *Convex Analysis*. Princeton University Press, Princeton, 1970.

[Roy88] H.L. Royden. *Real Analysis*. Macmillan, New York, 1988.

[Rud76] W. Rudin. *Principles of Mathematical Analysis*. McGraw-Hill, New York, 3rd edition, 1976.

Index

Printed in the United States
By Bookmasters